Viruses That Affect
the Immune System

ICN-UCI Conferences in Virology

Molecular Aspects of Picornavirus Infection and Detection
Edited by Bert L. Semler and Ellie Ehrenfeld

Common Mechanisms of Transformation by Small DNA Tumor Viruses
Edited by Luis P. Villarreal

Viruses That Affect the Immune System

Edited by

Hung Y. Fan
Cancer Research Institute, University of California, Irvine

Irvin S. Y. Chen
UCLA School of Medicine, Los Angeles, California

Naomi Rosenberg
Tufts University School of Medicine, Boston, Massachusetts

William Sugden
McArdle Laboratory, University of Wisconsin, Madison

American Society for Microbiology
Washington, D.C.

Copyright © 1991 American Society for Microbiology
1325 Massachusetts Ave., N.W.
Washington, DC 20005

Library of Congress Cataloging-in-Publication Data

Viruses that affect the immune system / edited by Hung Y. Fan . . . [et
al.].
     p. cm. — (ICN-UCI conference in virology)
   Papers presented at the 1990 ICN-UCI International Conference on
Virology, held in Newport Beach, Calif., in Mar. 1990.
   Includes index.
   ISBN 1-55581-032-2
   1. Retrovirus infections—Immunological aspects—Congresses.
2. Immune system—Infections—Congresses. 3. Herpesvirus diseases—
Immunological aspects—Congresses. I. Fan, Hung, 1947-
II. ICN Pharmaceuticals, Inc. III. University of California,
Irvine. IV. ICN-UCI International Conference on Virology (1990 :
Newport Beach, Calif.) V. Series.
   [DNLM: 1. HIV—immunology—congresses. 2. Immune System—
microbiology—congresses. 3. Viruses—immunology—congresses. QW
160 V8214 1990]
QR201.R47V57 1991
616.97'0194—dc20

CONTENTS

IV. ONCOGENESIS BY HERPESVIRUSES

CONTRIBUTORS

Yasmath F. Ahmed • Howard Hughes Medical Institute, Department of Medicine, and Department of Microbiology and Immunology, Duke University Medical Center, Durham, North Carolina 27710

Doug Aziz • Laboratory of Molecular Biology, Clinical Research Institute of Montreal, 110 Pine Avenue West, Montreal, Quebec, Canada H2W 1R7

Michael R. Bowman • BASF Bioresearch Corp., Cambridge, Massachusetts 02139

B. Kay Brightman • Department of Molecular Biology and Biochemistry and Cancer Research Institute, University of California, Irvine, California 92717

Steven J. Burakoff • Division of Pediatric Oncology, Dana-Farber Cancer Institute, and Department of Pediatrics, Harvard Medical School, Boston, Massachusetts 02115

David Camerini • Department of Microbiology and Immunology and Department of Medicine, University of California-Los Angeles School of Medicine, and Jonsson Comprehensive Cancer Center, Los Angeles, California 90024

Irvin S. Y. Chen • Department of Microbiology and Immunology and Department of Medicine, University of California-Los Angeles School of Medicine, and Jonsson Comprehensive Cancer Center, Los Angeles, California 90024

John M. Coffin • Department of Molecular Biology and Microbiology, Tufts University School of Medicine, 136 Harrison Avenue, Boston, Massachusetts 02111

Brian R. Davis • Medical Research Institute, 2200 Webster St., San Francisco, California 94115

David C. Diamond • Baxter Hyland Division, Duarte, California 91010

Harald Dinter • The Salk Institute for Biological Studies, La Jolla, California 92037

Hung Fan • Department of Molecular Biology and Biochemistry and Cancer Research Institute, University of California, Irvine, California 92717

Wayne N. Frankel • Department of Molecular Biology and Microbiology, Tufts University School of Medicine, 136 Harrison Avenue, Boston, Massachusetts 02111

Peter Ghazal • Department of Immunology, Research Institute of Scripps Clinic, La Jolla, California 92037

Warner C. Greene • Howard Hughes Medical Institute, Department of Medicine, and Department of Microbiology and Immunology, Duke University Medical Center, Durham, North Carolina 27710

Sarah M. Hanly • Howard Hughes Medical Institute, Department of Medicine, and Department of Microbiology and Immunology, Duke University Medical Center, Durham, North Carolina 27710

Zaher Hanna • Laboratory of Molecular Biology, Clinical Research Institute of Montreal, 110 Pine Avenue West, Montreal, Quebec, Canada H2W 1R7

Mark Hannink • McArdle Laboratory, University of Wisconsin, 1400 University Avenue, Madison, Wisconsin 53706

Wei-Shau Hu • McArdle Laboratory, University of Wisconsin, 1400 University Avenue, Madison, Wisconsin 53706

Ming Huang • Laboratory of Molecular Biology, Clinical Research Institute of Montreal, 110 Pine Avenue West, Montreal, Quebec, Canada H2W 1R7

Carlos Ibanez • Department of Immunology, Research Institute of Scripps Clinic, La Jolla, California 92037

Paul Jolicoeur • Laboratory of Molecular Biology, Clinical Research Institute of Montreal, 110 Pine Avenue West, Montreal, Quebec, Canada H2W 1R7, and Département de Microbiologie et d'Immunologie, Faculté de Médecine, Université de Montréal, Montreal, Quebec, Canada H3C 3J7

Katherine A. Jones • The Salk Institute for Biological Studies, La Jolla, California 92037

M. A. Kelliher • Department of Pathology, Department of Molecular Biology and Microbiology, and the Immunology Graduate Program, Tufts University School of Medicine, Boston, Massachusetts 02111

Nathaniel R. Landau • Department of Microbiology and Immunology, University of California, San Francisco, California 94143-0414

Qi-Xiang Li • Department of Molecular Biology and Biochemistry and Cancer Research Institute, University of California, Irvine, California 92717

Dan R. Littman • Department of Microbiology and Immunology, University of California, San Francisco, California 94143-0414, and Howard Hughes Medical Institute, University of California, San Francisco, California 94143-0724

J. McLaughlin • Department of Microbiology and Molecular Biology Institute, University of California at Los Angeles, Los Angeles, California 90024-1570

Kurtis D. MacFerrin • Department of Chemistry, Harvard University, Cambridge, Massachusetts 02138

Jay A. Nelson • Department of Immunology, Research Institute of Scripps Clinic, La Jolla, California 92037

Michael B. A. Oldstone • Division of Virology, Department of Neuropharmacology, Research Institute of Scripps Clinic, 10666 N. Torrey Pines Road, La Jolla, California 92037

Kathleen A. Page • Department of Microbiology and Immunology, University of California, San Francisco, California 94143-0414

Vinay K. Pathak • McArdle Laboratory, University of Wisconsin, 1400 University Avenue, Madison, Wisconsin 53706

Laurence T. Rimsky • Howard Hughes Medical Institute, Department of Medicine, and Department of Microbiology and Immunology, Duke University Medical Center, Durham, North Carolina 27710

Naomi Rosenberg • Department of Pathology, Department of Molecular Biology and Microbiology, and the Immunology Graduate Program, Tufts University School of Medicine, Boston, Massachusetts 02111

Stuart L. Schreiber • Department of Chemistry, Harvard University, Cambridge, Massachusetts 02138

Rachel Schrier • Department of Pathology, University of California at San Diego, La Jolla, California 92093

Philip L. Sheridan • The Salk Institute for Biological Studies, La Jolla, California 92037

Carole Simard • Laboratory of Molecular Biology, Clinical Research Institute of Montreal, 110 Pine Avenue West, Montreal, Quebec, Canada H2W 1R7

Samuel H. Speck • Dana-Farber Cancer Institute, Harvard Medical School, Boston, Massachusetts 02115

Jonathan P. Stoye • National Institute for Medical Research, The Ridgeway Mill Hill, London NW7 1AA, England

Bill Sugden • McArdle Laboratory for Cancer Research, University of Wisconsin, Madison, Wisconsin 53706

Howard M. Temin • McArdle Laboratory, University of Wisconsin, 1400 University Avenue, Madison, Wisconsin 53706

Marian L. Waterman • The Salk Institute for Biological Studies, La Jolla, California 92037

Clayton Wiley • Department of Pathology, University of California at San Diego, La Jolla, California 92093

O. N. Witte • Department of Microbiology and Molecular Biology Institute, University of California at Los Angeles, Los Angeles, California 90024-1570

Flossie Wong-Staal • University of California, San Diego, La Jolla, California 92093

FOREWORD AND ACKNOWLEDGMENTS

This volume represents papers presented at the 1990 ICN-UCI International Conference on Virology, "Viruses That Affect the Immune System," which was held in Newport Beach, Calif., in March, 1990. A number of human and animal viruses cause defects in the immune system, including neoplasias, immunodeficiency, and autoimmunity. Indeed, the emergence of human immunodeficiency virus (HIV) and the AIDS epidemic is the most dramatic recent example of this. The goal of the ICN-UCI International Conferences on Virology is to highlight timely areas of virology, with particular interest in medically relevant developments. Generous financial support from ICN Pharmaceuticals, Inc., made the meeting possible. Programmatic and administrative organization was carried out by the University of California Cancer Research Institute. The organizers are particularly indebted to Nita Driscoll of the Cancer Research Institute, University of California, Irvine, for providing staff support.

Viruses That Affect the Immune System
Edited by Hung Y. Fan et al.
© 1991 American Society for Microbiology, Washington, DC 20005

Chapter 1

Viruses That Affect the Immune System: an Overview of Retroviruses and Herpesviruses

Hung Fan, Irvin Chen, Naomi Rosenberg, and Bill Sugden

When humans or animals are infected by viruses, their immune systems are almost always affected. In most cases, the host immune system responds to viral infection by development of humoral or cell-mediated immune responses. In addition, for some viruses, infection can cause abnormalities of the immune system, including autoimmunity, neoplasias of immune cells, or immunodeficiency. Viruses that affect the immune system were the subject of the 1990 ICN-UCI International Conference on Virology, held in March, 1990, at Newport Beach, California. This volume contains papers presented at the symposium.

The immune system consists of various hematopoietic cells, as well as soluble factors in the blood and lymph (reviewed in reference 63). These cells and factors work together in a tightly regulated fashion to carry out the different functions involved in immunity. The tight regulation is important since it allows rapid immunological responses (e.g., to infections), while at the same time avoiding undesirable or unnecessary immune activities (e.g., autoimmunity). Immune system cells include lymphocytes (B- and T-lymphocytes), which respond to specific antigens, as well as cells that carry out effector functions, such as phagocytes (monocyte/macrophages and neutrophilic granulocytes), mast cells, and eosinophils. In terms of lymphocytes, B-

Hung Fan • Department of Molecular Biology and Biochemistry and Cancer Research Institute, University of California, Irvine, California 92717. **Irvin Chen** • Department of Medicine, UCLA School of Medicine, Los Angeles, California 90024. **Naomi Rosenberg** • Department of Pathology, Tufts University School of Medicine, Boston, Massachusetts 02111. **Bill Sugden** • McArdle Laboratory, University of Wisconsin, Madison, Wisconsin 53706.

lymphocytes are responsible for production of soluble antibodies (humoral immunity), while T-lymphocytes are responsible for cell-mediated immunity. T-lymphocytes can be subdivided further into cytotoxic (or killer) and helper cells. Cytotoxic T-lymphocytes bind and directly kill target (e.g., virus-infected) cells in an antigen-dependent manner, while T-helper lymphocytes provide "help" in maturation of both B- and T-lymphocytes. Two characteristic surface proteins that generally distinguish helper and cytotoxic T-lymphocytes are CD4 (present on helper T-cells) and CD8 (present on cytotoxic T-cells) (7, 62). T-helper lymphocytes play a pivotal role in immune system function since they are important for development of both humoral and cell-mediated immune responses.

Soluble factors of the immune system include effector molecules such as antibodies, produced by the plasma cells (the differentiated elements of the B-cell lineage), and components of the complement system, produced in part by hepatocytes. Cells of the immune system also produce a large number of growth regulatory molecules that act on the immune cells themselves. T-lymphocytes produce interleukin-1 (IL-1), IL-2, IL-3, IL-4, IL-5, IL-6, and IL-7, along with gamma interferon (reviewed by Gillis [24]). Macrophages also produce IL-1 and IL-6, tumor necrosis factor, tumor growth factor beta, alpha and beta interferons, and other growth regulatory molecules (reviewed in reference 19). Each of these soluble mediators affects multiple cell types in a variety of ways. For example, IL-2 stimulates T-lymphocyte proliferation and induces production of other lymphokines by T-cells (21, 81), it increases natural killer cell activity (27), and it stimulates proliferation and differentiation of B-lymphocytes under appropriate conditions (55, 87). Some mediators such as IL-1 and tumor necrosis factor also play a role in the normal inflammatory response and in inducing fever and cachexia (17, 18).

In terms of viral effects on the immune system, viruses can lyse particular subsets of immune cells, alter their functions, or induce malignant transformation. Thus these viruses can cause autoimmunity (e.g., lymphocytic choriomeningitis virus, discussed in the chapter by Oldstone in this volume), lymphoid or myeloid malignancies (e.g., retroviruses and Epstein-Barr virus), or immunodeficiencies (e.g., retroviruses and cytomegalovirus). A general feature of viruses that affect the immune system is that they can establish infection (productive, latent, or both) in immune system cells. Thus, understanding the replication of these viruses, particularly in cells of the immune system, is important to elucidating the mechanisms by which they cause immunological damage.

Many of the viruses that cause immune system abnormalities are retroviruses or herpesviruses. Therefore a brief overview of these viruses will be presented in the context of their pathogenic effects.

RETROVIRUSES

Structure

Retroviruses are enveloped positive-stranded RNA viruses (reviewed in reference 79). Each retrovirus particle contains two copies of genomic RNA (8 to 10 kb) held together by hydrogen bonding in a dimeric structure (Fig. 1). Retrovirus virions contain an inner core consisting of the genomic RNA and associated cellular tRNAs, capsid proteins (M, CA, p12, and NC: products of the *gag* gene), and enzymes (PR, RT, and IN: products of the *pol* gene) (45). (For avian retroviruses, protease [PR] is a product of the *gag* gene.) The core is surrounded by an envelope consisting of cell-derived lipid bilayer and two envelope proteins (TM and SU: products of the *env* gene). The order of the genes on the viral genome is *gag, pol, env* from 5' to 3' (see Fig. 1). The human retroviruses human T-cell leukemia virus (HTLV) and

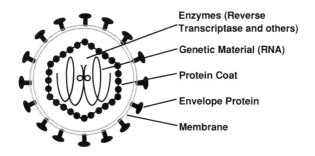

The RNA Genetic Material

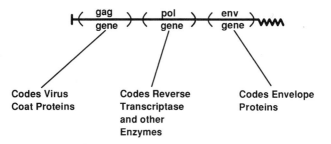

Figure 1. Retroviral structure. The structure of a retrovirus particle is shown in the top of the figure. The bottom of the figure shows the genetic organization of viral RNA. Standard retroviruses contain three genes, *gag, pol,* and *env,* which encode the proteins indicated. Viral RNA is capped at the 5' end and polyadenylated at the 3' end, like cellular mRNAs.

human immunodeficiency virus (HIV) belong to virus classes that have extra genes in addition to *gag, pol,* and *env,* as described below.

Replication

An outline of a retroviral replication cycle is shown in Fig. 2 (and reviewed in reference 79). Virus particles bind to a specific receptor at the cell surface and are internalized into the cytoplasm, and the envelope is removed. Within the internalized core, virion reverse transcriptase uses viral RNA as a template to synthesize linear double-stranded DNA. The double-stranded DNA is transported to the nucleus (perhaps as a nucleoprotein complex [4]), where it is integrated into host chromosomal DNA at multiple sites to form the provirus. Proviral DNA is then recognized by cellular RNA polymerase II, which transcribes the provirus to give a viral RNA molecule identical to genomic RNA. The viral RNA is transported to the cytoplasm along two

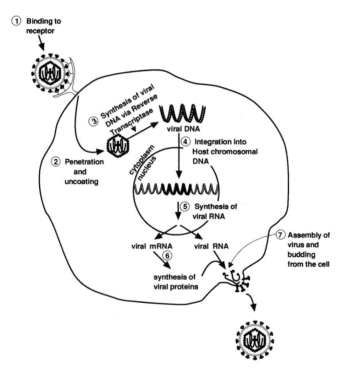

Figure 2. The retroviral life cycle. The replication cycle for a typical retrovirus is shown. The events occur temporally in the order indicated, starting with binding to a cellular receptor and ending with assembly of virus and budding from the cell.

pathways. One pathway (with and without splicing) yields viral mRNA, which is translated to form precursor polyproteins for the three viral genes. The second pathway yields unspliced genomic RNA, which is packaged with viral proteins into virus particles that bud from the cell surface. In general, retrovirus infection is not lytic to the cell, although HIV is an exception. Virus-infected cells continue to harbor provirus and shed infectious virus.

Viral DNA synthesized by reverse transcriptase is longer than the template RNA, due to the presence of long terminal repeats (LTRs) at either end (Fig. 3) (36, 37). The LTRs can be subdivided into three regions, termed U3,

A. Retroviral DNA

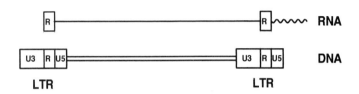

B. A retroviral LTR (M-MuLV)

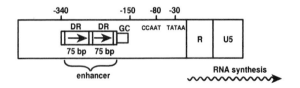

C. Enhancer core sequences in the M-MuLV Direct Repeat

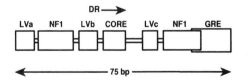

Figure 3. Organization of a retroviral LTR. The organization of a retroviral LTR is shown. Relationship of LTR sequences to the viral RNA is shown in part A. Part B shows DNA sequence elements within the LTR of Moloney MuLV (M-MuLV), including promoter-proximal elements and enhancers. Part C shows the enhancer core motifs present in each 75-bp direct repeat of the M-MuLV enhancer. (From reference 20.)

R, and U5. When proviral DNA is transcribed into viral RNA, transcription begins at the U3-R junction of the upstream LTR, and cleavage-polyadenylation occurs at the R-U5 junction in the downstream LTR. Nucleotide sequences in the U3 region are particularly important for viral transcription. They include typical RNA polymerase II promoter elements (TATA and CCAAT boxes) as well as transcriptional enhancers (Fig. 3). The enhancers are of particular interest since (like many cellular enhancers) they show cell-type preference for activity and they influence the pathogenic behaviors of several retroviruses (20).

Retroviruses generally only efficiently infect cells that are undergoing cell division (84). In adult individuals, hematopoietic cells are one of relatively few cell types for which cell division takes place. Thus, the frequent association of retrovirus pathogenicity with the immune system may result from these two facts.

HTLV-I and -II belong to a class of retroviruses (along with bovine leukemia virus) that have somewhat more complex genomes (6). In addition to the *gag, pol,* and *env* genes typical of all retroviruses, HTLV has an additional region originally termed "X" (69). The X region is expressed by alternative translation of a doubly spliced mRNA into two (or perhaps three) proteins including *tax* and *rex* (6). A diagram of the HTLV-I or -II genome is shown in Fig. 4. Both *tax* and *rex* encode regulatory proteins for viral expression. The *tax* protein is a transcriptional transactivator of the HTLV LTR that up-regulates transcription (71). Repeated sequences (21-bp repeats) in the U3 regions of the HTLV LTR are required for *tax* response (56), although *tax* protein does not directly bind to them (53). Instead, cellular factors bind to the 21-bp repeats; thus *tax* may act by directly or indirectly interacting with these cellular 21-bp binding proteins. The *rex* protein is a regulatory protein that shifts the balance of viral mRNAs from doubly spliced mRNAs (encoding X region products) to unspliced or singly spliced mRNAs (encoding virion proteins) (28, 33, 66). Thus, *rex* provides HTLV with a temporal switch for viral expression from an "early" pattern, characterized by high levels of X region products, to a "late" pattern characterized by high

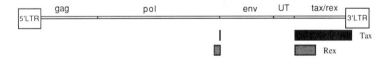

Figure 4. The HTLV genome. The proviral genome of HTLV-I/-II is shown schematically with the prototypical *gag, pol,* and *env* genes indicated. The *tax/rex* genes, unique to HTLV, are shown at the 3' end. The coding regions for *tax* and *rex* are indicated below the proviral genome. The 5' and 3' LTRs are indicated at the ends of the proviral genome. "UT" refers to an untranslated region, which has as yet no known function and does not appear to encode a protein.

levels of virion proteins. The mechanism of *rex* action has been of considerable recent interest; it appears to involve mRNA transport as well as other steps. *rex* protein response sequences are found at the 3' end of viral mRNA in the R region. The chapter by Greene et al. in this volume discusses HTLV-I *rex* action.

HIV is a member of the lentivirus class of retroviruses. HIV has an even more complicated genome than HTLV, with at least six regulatory proteins encoded. A diagram of the HIV genome is shown in Fig. 5. Interestingly, HIV encodes two regulatory proteins that are functionally analogous to the HTLV regulatory proteins *tat* and *rev* (13, 32, 82). HIV *tat* (like HTLV *tax*) is a positive transactivator of viral expression (72), although its mechanism of action is different from that of *tax*. Sequences that confer *tat* responsiveness (TAR sequences) are located in the R region of the HTLV LTR (61) and must be expressed as mRNA (59). While *tat* does not directly bind to TAR DNA sequences, other cellular factors do (83; see Dinter et al., this volume). On the other hand, *tat* protein may bind TAR RNA (see Wong-Staal, this volume). The current view is that *tat* protein interacts with cellular factors at or near the transcription initiation complex. HIV *rev* has a function analogous to that of HTLV *rex* in shifting the balance of viral mRNAs toward mRNAs for virion proteins (49); HTLV *rex* can actually partially substitute for HIV *rev* function (49). The *rev* protein also binds to specific sequences in viral RNA, although in this case the sequences are in the *env* region (32, 49, 85). The chapter by Wong-Staal in this volume discusses mechanisms of HIV *tat* and *rev* action.

Tumorigenesis

In terms of tumorigenesis, retroviruses can be divided into two broad classes: acute transforming and nonacute retroviruses (76). Acute transforming retroviruses cause rapid neoplasms and frequently transform cells in culture. These properties are due to the fact that they carry oncogenes. Viral oncogenes are derived from normal cell proto-oncogenes by retroviral capture (3); during this process various alterations in the proto-oncogene have resulted, including single-base mutations, deletions, and substitutions (10).

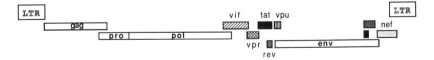

Figure 5. HIV-1 genetic structure. The HIV-1 proviral genome is shown schematically. The *gag, pol,* and *env* genes typical for all retroviruses are indicated. Other genes, *vif, vpr, rev, tat, vpu,* and *nef,* are unique to lentiviruses of the HIV-1 type.

These alterations are important in the oncogenic potential of viral oncogenes. Study of acute transforming retroviruses originally led to the identification of cellular proto-oncogenes (3); subsequent experiments indicated that in the normal cell, proto-oncogenes are generally involved in some aspect of gene regulation or growth control. Other studies also revealed that alterations or activations of proto-oncogenes are frequently one of the steps involved in development of nonviral and spontaneous tumors.

A number of acute transforming retroviruses cause tumors of the immune system (leukemias), notably viruses carrying the *myc* (25), *abl* (65), and *rel* (73) oncogenes. Temin et al. (this volume) discuss the *rel* oncogene (from the avian virus REV-T). Normal cell *rel* protein is a transcriptional transactivator that shares homology with NF-κB transcription factor (23, 41), and alterations in the viral oncogene form (v-*rel*) confer greatly increased activity. Kelliher et al. (this volume) discuss the *abl* oncogene. Like several other viral oncogenes, v-*abl* protein is a tyrosine-specific protein kinase, with the viral protein having markedly higher activity than the c-*abl* protein. The *abl* oncogene is also particularly interesting, because chromosomal translocation of the c-*abl* proto-oncogene to another specific chromosomal location (*bcr*) is characteristic (the Philadelphia translocation) in human chronic myelogenous leukemia and in some human acute lymphoblastic lymphomas (reviewed in reference 61). The chapter by Kelliher et al. describes in vivo tumorigenic potentials of retroviruses carrying the v-*abl* and *bcr*-c-*abl* oncogenes.

HTLV-I causes adult T-cell leukemia, particularly in high endemic areas for the virus such as Japan and the Caribbean (6). However, even among HTLV-I-infected individuals the progression rate to leukemia is quite slow (6). This contrasts with the relatively rapid course of disease for acute transforming retroviruses in animal models. On the other hand, hallmarks of leukemogenesis by nonacute retroviruses (common sites of integration in independent tumors; see below) are not found in HTLV-I-induced adult T-cell leukemia tumors (68). The most likely explanation of HTLV-I leukemogenesis involves *tax* protein. This protein also transcriptionally transactivates selected cellular genes, including the IL-2 and IL-2 receptor genes (12). Thus, HTLV-I infection of a T-lymphocyte could result in abnormally high production of both IL-2 and its receptor in the same cell, resulting in an autocrine loop and continual growth stimulation. In support of this, T-lymphocytes can be immortalized by HTLV-I or -II infection (6). Presumably, HTLV infection provides one step (via *tax*) in a multistep tumorigenic process, while the other steps occur by nonviral means. This could also account for the long time lag between viral infection and development of leukemia.

Nonacute retroviruses differ from acute transforming retroviruses in that they are standard replication-competent retroviruses and they lack oncogenes

(76). The time course for disease induction is generally slower than for acute transforming retroviruses. The important primary mechanism for nonacute retrovirus pathogenesis was first identified by Hayward et al. (31) in studies on avian leukosis virus-induced B-lymphoid tumors. They showed that independent avian leukosis virus-induced tumors all had proviral insertions adjacent to the c-*myc* proto-oncogene and that a readthrough transcript from the downstream LTR into the c-*myc* sequences was present. The net result was overexpression of c-*myc* in the tumors. This mechanism has been termed "promoter insertion," since the viral LTR promoter is inserted in tumor cells upstream of the c-*myc* gene and leads to abnormally high expression under LTR control (Fig. 6). The promoter insertion mechanism also provided an explanation for at least part of the long latency of nonacute retroviruses. As mentioned above, during infection proviral insertion occurs at multiple (virtually random) sites in different cells, and the probability of insertion next to a particular location (such as c-*myc*) is quite low. Thus multiple rounds of infection in an animal may be required before a single cell sustains a

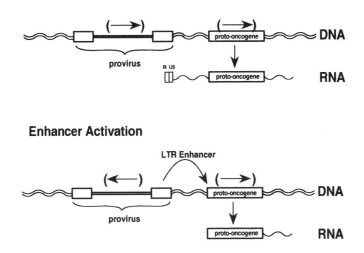

Figure 6. LTR activation of proto-oncogenes. In tumors induced by nonacute retroviruses, proviral insertion next to cellular proto-oncogenes is an important mechanism in tumorigenesis. Two related mechanisms are indicated. In classical promoter insertion, a readthrough transcript initiating in the downstream LTR and continuing into the proto-oncogene results in high expression of a chimeric retroviral proto-oncogene transcript. In enhancer activation, the enhancer in the retroviral LTR activates the proto-oncogene's own promoter, leading to overexpression of the normal transcript. (From reference 20.)

proviral integration next to c-*myc*; once this happens, that cell may then grow out and form the tumor.

The promoter insertion mechanism of leukemogenesis was subsequently expanded to a more general mechanism. In other retroviral systems, independent tumors also showed common sites of proviral insertion next to proto-oncogenes and overexpression of the proto-oncogenes (14, 52, 57). However, the orientations of the provirus and the proto-oncogene were inappropriate for classical promoter insertion. In these cases the enhancer sequences in the viral LTR activate the proto-oncogene's own promoter, leading to overexpression (Fig. 6). Thus, the generalized mechanism can be referred to as "LTR activation" of proto-oncogenes, including both classical promoter insertion and enhancer activation.

Different nonacute retroviruses induce neoplasms by activation of a variety of proto-oncogenes. Indeed, several proto-oncogenes were identified on the basis of LTR activation in tumors (e.g., *int-1* for murine mammary tumor virus [52] and *pim-1* for Moloney murine leukemia virus [MuLV] [14]). Currently, more than 20 different proto-oncogenes have been implicated in retroviral LTR activation. For some virus-induced tumors, LTR activation may involve a single proto-oncogene, e.g., c-*myc* in avian leukosis virus-induced B-lymphomas (31, 57). For others, LTR activation may involve multiple proto-oncogenes or insertion sites, used in either-or or additive fashion. For instance, Moloney MuLV-induced T-lymphomas in mice involve four predominant proto-oncogenes or insertion sites (c-*myc, pim-1, pim-2,* and *pvt-1* [14, 26, 46, 70]), and Moloney MuLV-induced T-lymphomas in rats may involve three or four additional loci (78, 80).

For nonacute retroviruses, the enhancers in the LTRs are important determinants in pathogenic potential (reviewed in reference 20). This can be rationalized in terms of the LTR activation mechanism for leukemogenesis. In either classical promoter insertion or enhancer activation, there is a requirement for an LTR enhancer that is highly active in the target cell type. Thus, when Friend MuLV (which induces erythroleukemia) and Moloney MuLV (which induces T-lymphoid leukemia) were compared by generation of molecular recombinants, the specificity of disease was mapped to the enhancer regions of the respective viruses (8). Indeed, the Friend MuLV enhancer is preferentially active in erythroid cells in comparison to lymphoid cells, and the opposite is true for the Moloney MuLV enhancer. Likewise, when the enhancer region of a virus was weakened (e.g., by deletion of one copy of a tandemly repeated enhancer), the time course of disease was significantly lengthened (35, 47). Other alterations in enhancer regions of nonacute retroviruses have provided insights into the leukemogenic process as well. Fan et al. (this volume; chapter 11) describe studies using an enhancer variant of Moloney MuLV.

For some nonacute retroviruses, genetic exchange with endogenous retroviral genes may be important in pathogenesis. This is particularly true for murine retroviruses such as MuLV. Normal mice contain approximately 40 to 60 copies of endogenous MuLV-related proviruses in their genomes, and some of these can recombine with exogenous MuLVs during infection (75). With regard to leukemogenesis, *env* gene recombinants have been of particular interest. Hartley et al. (30) described the appearance of "mink cell focus-inducing" or MCF recombinants during MuLV leukemogenesis and suggested that these MCF recombinants may play a role in the leukemogenic process. MCF derivatives are *env* recombinants involving endogenous (polytropic) MuLV-related proviruses that encode envelope glycoprotein (gp70 or SU) that interacts with a different cellular receptor from that of MuLVs (ecotropic) (48, 77). The appearance of MCF recombinants late in the disease process, as well as molecular evidence for their presence in resultant tumors, has led some to suggest that MCF recombinants are the "proximal leukemogens" (30). Fan et al. (this volume, chapter 11) suggest that MCF recombinants may also be involved early in the leukemogenic process for Moloney MuLV. Coffin et al. (this volume) describe a systematic molecular analysis of endogenous murine retroviruses and their contributions to the leukemogenic process.

As indicated above, leukemogenesis by nonacute retroviruses has a slow time course. At least part of the long latency can be ascribed to the LTR activation mechanism, as also discussed above. However, as for virtually all carcinogenic processes, tumorigenesis is probably multistep. While nonacute retroviruses carry out at least one step, LTR activation of proto-oncogenes, it is possible that they may carry out additional steps as well. Fan et al. (this volume; chapter 11) discuss the possibility that Moloney MuLV also induces early events in leukemogenesis, likely through MCF recombinants.

Immunodeficiency

Acquired immune deficiency (AIDS) and its etiological agent, HIV, are of great clinical importance today. There are actually two related strains of HIV, HIV-1 and HIV-2 (82). HIV-1 is the predominant virus associated with the current AIDS epidemic and is found in sub-Saharan Africa, North America, Europe, and other parts of the world. HIV-2 is currently found predominantly in western coastal Africa, with little presence in North America or Europe. There is some suggestion that HIV-2 may cause AIDS less efficiently than does HIV-1 (82).

The time course for disease after HIV-1 infection is noteworthy. Infected individuals generally show few symptoms for extended periods of time; current estimates are that on the average 8 to 10 years elapse between initial infection and development of clinical AIDS (82). Early during the asymptomatic period, most individuals develop circulating antibodies to HIV, the common diagnostic test for infection. With the exception of transitory viremia immediately after infection, infectious virus is generally difficult to recover from infected asymptomatic individuals, and the number of infected cells in the blood is relatively low (typically less than 1 in 10^5 [29]). As the disease progresses, the immune systems of infected individuals become impaired, heralded by a precipitous decline in the number of CD4[+] T-helper lymphocytes. As described above, T-helper lymphocytes are important for both humoral and cellular immunity, so depletion of these cells ultimately leads to profound immunological deficiency. The major clinical problems associated with AIDS result from immunological failure and include opportunistic infections (e.g., *Pneumocystis carinii* pneumonia and cytomegalovirus retinitis) and cancers (e.g., lymphomas and Kaposi's sarcoma). In addition, AIDS-associated neurological symptoms (e.g., dementias) may secondarily result from release within the brain of biologically active molecules (cytokines) from HIV-infected cells (perhaps macrophages and endothelial cells) (50).

In terms of understanding the mechanisms of HIV pathogenesis, certain aspects of HIV replication are important. First, the cellular receptor for HIV is the CD4 surface molecule (67). As described above, CD4 is a characteristic surface protein of T-helper lymphocytes, and it is also present on some monocyte/macrophages. Thus, the major targets in the body for HIV infection are these two cell types. Interactions of the HIV envelope protein SU (or gp120) with CD4 protein are discussed by Camerini and Chen, Diamond et al., and Page et al. (this volume). These authors define critical regions of the molecules necessary for binding interactions. Diamond et al. also describe effects of HIV gp120 binding on T-helper lymphocyte function.

An important feature of HIV is that infection of CD4[+] lymphocytes is lytic (22), an unusual result when compared with most other retroviruses. This suggests a direct infection mechanism for the eventual loss of T-helper lymphocytes as HIV infection progresses to AIDS. However, it is unclear whether T-helper lymphocyte depletion results exclusively from lytic infection of T-helper lymphocytes or whether other indirect mechanisms of depletion may also play a role (22). On the other hand, HIV infection of monocytes/macrophages is generally nonlytic. This also has several important consequences. First, these cells may act as reservoirs for HIV infection in the infected individual, since they are not lysed. Second, monocytes/macrophages are the sources of several powerful cytokines, and infected cells might produce abnormal amounts of these factors (see above).

Another important feature of HIV infection is that the virus persists over prolonged periods of time in infected individuals, even in the presence of humoral (and perhaps cell-mediated) immune responses (34). Thus HIV can establish persistent and/or latent states in infected individuals. In vitro evidence for latent HIV infections in both monocytes and T-lymphoid cells has been reported (9, 60). Another feature of HIV infection is that the virus shows high rates of mutation in infected individuals, manifested as changes in the *env* glycoproteins. Such alterations may contribute to the ability of the virus to evade the host immunological response, since *env* SU protein (gp120) is a predominant viral target for neutralizing antibody. *env* gene alterations may also be important in altering the host range of HIV towards cells such as macrophages.

The extent of HIV infection in target T-helper lymphocytes and monocytes is an important consideration. In fact, in asymptomatic individuals, the frequency of infected cells is quite low (see above). One contributory factor to the low frequency of infection may be the requirement of active cell division for establishment of productive infection. Studies on HIV reverse transcription in resting and growing cells show that cells must be cycling in order for reverse transcription to be completed (84). It should be noted that the majority of circulating lymphocytes and monocytes are not dividing; thus the low frequency of productively infected cells in asymptomatic people is not completely surprising.

Animal model systems for studying retroviral immunodeficiency will be important to understanding the mechanisms involved. Currently, several are under investigation, including macaque simian immunodeficiency virus in rhesus macaques (16, 39), feline immunodeficiency virus (58) and feline leukemia virus (Sarma-C) (54) in cats, and the Duplan strain of MuLV in mice (MAIDS [2, 5]). A study of the MAIDS model system is described by Jolicoeur et al. (this volume). In the MAIDS system, immunodeficiency is caused by a defective viral genome. Interestingly, the immunological defects appear to result from neoplastic transformation of lymphoid cells.

HERPESVIRUSES

Herpesviruses resemble retroviruses in one way: they are diverse. In most other respects, herpesviruses represent the antipodes of retroviruses. In particular, their nucleic acid is composed of linear duplex DNA molecules that range from slightly more than 100 kbp to about 240 kbp. Productive infection by cells with herpesviruses always leads to death of the cell from which the virus is released. In addition to infecting cells productively, some herpesviruses also infect cells latently. In these instances, the viral DNA is maintained in the cell, but only a small subset of viral information is expressed. In this collection of reports, two members of the herpesvirus family that are human pathogens are considered: human cytomegalovirus and Epstein-Barr virus.

Human cytomegalovirus infects human fibroblast strains in cell culture. Much of what is known about the regulation of viral gene expression and genome replication comes from studies of virus infection in this cell type (reviewed by Stinski [74]). Such studies have shown that viral RNA is expressed in three temporal groups, immediate early, early, and late. The latter two groups are in part regulated by products of the first group. Early genes encode proteins involved in viral DNA replication, and late genes encode viral proteins that participate in the assembly of virions. The regulation of one family of immediate early genes expressed from the major immediate early promoter has been a focus of studies on human cytomegalovirus (Nelson et al., this volume). A plethora of *cis*-acting DNA sequences have been shown to negatively and positively affect transcription from the major immediate early promoter by binding different transcription factors of the host cell. Different levels of these factors in different types of cells can affect the level of expression of the major immediate early promoter and thereby influence the course of infection by the herpesvirus (Nelson et al., this volume).

Viral DNA synthesis for human cytomegalovirus is only beginning to be studied now. Presumably this virus synthesizes its DNA using a number of viral gene products following a rolling-circle mechanism, as has been suggested for herpes simplex virus type 1 (38). During the course of synthesis, the viral DNA is amplified 100- to 1,000-fold and is eventually packaged into virions which are enveloped and released from the cell, accompanied by cell death.

In vivo, human cytomegalovirus infects cell types other than fibroblasts (reviewed in reference 1). There is evidence that it infects epithelial cells and cells of the hemopoietic system; fetal infection with human cytomegalovirus has been associated with abnormalities in the development of the central nervous system. It is also clear that the pathogenesis of human cytomegalovirus is affected not only by the type of cell infected but also by the state of the immune response of the host (Nelson et al., this volume). In individuals who are immunocompromised, because they have been treated with immunosuppressive drugs or because they have AIDS, recurrent cytomegalovirus infection poses a major health problem.

Epstein-Barr virus in vivo infects lymphoid cells and epithelial cells. In vitro it efficiently infects only human B-lymphocytes, and its infection of this cell type has been studied in cell culture (reviewed in reference 40). In contrast to human cytomegalovirus, Epstein-Barr virus infects cells in culture latently. That is, when this virus infects a resting human B-lymphocyte, the outcome of that infection initially is not a productive cycle of viral replication, but rather stimulation of the B-cell to divide and to become immortalized. In the immortalized cell, Epstein-Barr viral DNA is maintained as a plasmid using a specific *cis*-acting element as an origin of plasmid replication termed *ori*P. A subset of viral genes is expressed in the immortalized cell, which is described as being latently infected. The subset of latently expressed genes includes approximately 10 genes whose transcription is complex and often involves multiply spliced mRNA molecules (Speck, this volume). These viral genes are thought to affect expression of cellular genes such that they act in concert to induce the cell to divide and maintain it in the dividing state (Sugden, this volume). Some of the viral genes also are involved in affecting viral gene expression, and one, EBNA-1, is required to maintain the viral DNA as a plasmid in the immortalized cell. In immortalized cells, a rare event happens such that a particular viral gene, termed BZLF-1, is expressed. Expression of this gene appears to be sufficient to induce the lytic phase of the viral life cycle (11). Once the BZLF-1 gene is induced, two temporal sets of viral genes are expressed which may be likened to the early and late genes of human cytomegalovirus. These genes compose the rest of the Epstein-Barr virus genome. They encode the proteins required for viral DNA replication during the lytic phase of the viral life cycle and for formation of the virion. DNA synthesis by Epstein-Barr virus during the lytic phase of its life cycle uses a rolling-circle mechanism to increase the viral DNA approximately 100- to 1,000-fold. The viral DNA is packaged and released upon lysis of the cell. The frequency of activation of the lytic cycle in clones of cells immortalized by Epstein-Barr virus varies. It ranges from 1 cell per 100 to less than 1 cell per 10^6 per cell cycle. In all cases, the spontaneous activation of the lytic cycle is inefficient. The study of Epstein-Barr virus is therefore difficult because of the lack of a genuine lytic host.

In most of the world, in vivo infection of B-lymphocytes in people by Epstein-Barr virus leads either to no detectable disease or to infectious mononucleosis (reviewed by Miller [51]). In the latter case, B-cells proliferate, induce a specific T-cell-mediated cytotoxic response, and eventually are limited in their proliferation by this T-cell response such that few cells harboring virus remain in the host. These cells apparently are both a subset of B-lymphocytes and, perhaps, epithelial cells within the salivary gland.

Infection of some epithelial cells, particularly in people in China and Alaska, is associated with nasopharyngeal carcinoma, long after primary infection. Viral DNA is found in the tumor cells (42), and prospective seroepidemiologic surveys indicate that Epstein-Barr virus is likely to be causally associated with this disease (86). Little more is known about the contribution of the virus or of the host's immune response to this particular neoplasm. In other regions of the world Epstein-Barr virus infection of B-lymphocytes in some youngsters is causally associated with Burkitt's lymphoma. In these instances, holoendemic malaria is likely to be a cofactor for the lymphoma.

Viral infection occurs 9 to 48 months before detection of the lymphoma (15), and the tumor cells almost always have chromosomal translocations that juxtapose an immunoglobulin locus to the c-*myc* locus (43). The role of Epstein-Barr virus in Burkitt's lymphoma presumably is to induce infected B-cells to divide and to become immortalized, thereby providing a proliferating population of B-cells in which a rare chromosomal translocation can occur. The role of chronic malaria is as an immunosuppressive agent. It has been shown that during the acute phase of malarial infection, youngsters lack a vigorous T-cell-mediated cytotoxic response to their own Epstein-Barr virus-immortalized cells (44). Burkitt's lymphoma, therefore, represents the unfortunate confluence of a particular kind of immunosuppression mediated by malaria, a particular kind of cell proliferation mediated by Epstein-Barr virus, and a particular kind of genetic rearrangement that occurs rarely and only in B- or T-lymphocytes.

As with human cytomegalovirus, the immune status of the host plays an important role in determining the outcome of infection by Epstein-Barr virus. Patients who are immunocompromised because they are treated with immunosuppressive drugs or because they have AIDS are prone to develop Epstein-Barr virus-associated lymphoid tumors akin to Burkitt's lymphoma. In particular, bone marrow transplants, heart transplant recipients, and renal transplant recipients can develop B-cell lymphomas. AIDS patients also can develop B-cell lymphomas associated with Epstein-Barr virus. These observations support the contention that a person's immune response to infection by Epstein-Barr virus is crucial in determining whether the outcome of the infection will be benign or malignant.

REFERENCES

1. **Alford, C. A., and W. J. Britt.** 1985. Cytomegalovirus, p. 1981–2010. *In* B. N. Fields and D. M. Knipe (ed.), *Virology.* Raven Press, New York.
2. **Aziz, D. C., Z. Hanna, and P. Jolicoeur.** 1989. Severe immunodeficiency disease induced by a defective murine leukemia virus. *Nature* (London) **338:**505–508.
3. **Bishop, J. M., and H.E. Varmus.** 1982. Functions and origins of retroviral transforming genes, p. 999–1108. *In* R. Weiss, N. Teich, H. Varmus, and J. Coffin (ed.), *RNA Tumor Viruses. Molecular Biology of Tumor Viruses,* 2nd ed. Cold Spring Harbor Laboratory, Cold Spring Harbor, N.Y.
4. **Bowerman, B., P. O. Brown, J. M. Bishop, and H. E. Varmus.** 1989. A nucleoprotein complex mediates the integration of retroviral DNA. *Genes Dev.* **3:**469–478.
5. **Buller, R. M., R. A. Yetter, T. N. Frederickson, and H. C. Morse, 3d.** 1987. Abrogation of resistance to severe mousepox in C57Bl/6 mice infected with LP-BM5 murine leukemia viruses. *J. Virol.* **61:**383–387.
6. **Cann, A. J., and I. S. Y. Chen.** 1990. HTLV-8 and -II, p. 1501–1528. *In* B. N. Fields, D. M. Knipe, R. M. Chanock, M. S. Hirsch, J. L. Melnick, T. P. Monath, and B. Roizman (ed.), *Virology,* 2nd ed. Raven Press, New York.
7. **Cantor, H., and E. A. Boyse.** 1975. Functional subclasses of T lymphocytes bearing different

Lyt antigens. II. Cooperation between subclasses of Lyt cells in the generation of killer activity. *J. Exp. Med.* **141**:1390–1399.

8. **Chatis, P. A., C. A. Holland, J. W. Hartley, W. P. Rowe, and N. Hopkins.** 1983. Role for the 3′ end of the genome in determining disease specificity of Friend and Moloney murine leukemia virus. *Proc. Natl. Acad. Sci. USA* **80**:4408–4411.

9. **Clouse, K.A., D. Powell, I. Washington, G. Poi, K. Strebel, W. Farrar, B. Barnstad, J. Kovacs, A. S. Fauci, and T. M. Folks.** 1989. Monokine regulation of HIV-1 expression in a chronically-infected human T cell clone. *J. Immunol.* **142**:431–438.

10. **Cooper, G. M.** 1990. *Oncogenes.* Jones & Bartlett, Boston.

11. **Countryman, J. U., and G. Miller.** 1985. Activation of expression of latent Epstein-Barr herpesvirus after gene transfer with a small cloned subfragment of heterogeneous viral DNA. *Proc. Natl. Acad. Sci. USA* **82**:4085–4089.

12. **Cross, S. L., M. B. Feinberg, J. B. Wolf, N. J. Holbrook, F. Wong-Staal, and W. J. Leonard.** 1987. Regulation of the human interleukin-2 receptor alpha promoter: activation of a non-functional promoter by the transactivator gene of HTLV-I. *Cell* **49**:47–56.

13. **Cullen, B. R., and W. C. Greene.** 1989. Regulatory pathways governing HIV-1 replication. *Cell* **58**:423–426.

14. **Cuypers, H. T., G. Selten, W. Quint, M. Ziljstra, E. R. Maandag, W. Boelens, P. van Wezenbeek, C. Melieft, and A. Berns.** 1984. Murine leukemia virus T-cell lymphomagenesis: integration of proviruses in a distinct chromosomal region. *Cell* **37**:141–150.

15. **de-The, G., A. Geser, N. E. Day, P. M. Tukei, E. H. William, D. P. Beri, P. G. Smith, A. G. Dean, G. W. Bornkamm, P. Feorino, and W. Henle.** 1978. Epidemological evidence for causal relationship between Epstein-Barr virus and Burkitt's lymphoma from Uganda prospective study. *Nature* (London) **274**:756–761.

16. **Dewhurst, S., J. E. Embretson, D. C. Anderson, J. I. Mullins, and P. N. Fultz.** 1990. Sequence analysis and acute pathogenicity of molecularly cloned SIV. *Nature* (London) **345**:636–640.

17. **Dinarello, C.A., J. G. Cannon, S. M. Wolff, H. A. Bernheim, B. Beutler, A. Cerami, I. S. Figari, M. A. Palladino, and J. V. O'Connor.** 1986. Tumor necrosis factor (cachectin) is an endogenous pyrogen and induces production of interleukin 1. *J. Exp. Med.* **163**:1433–1450.

18. **Duff, G. W., and S. K. Durum.** 1983. The pyrogenic and mitogenic actions of interleukin-1 are related. *Nature* (London) **304**:449–451.

19. **Durum, S. K., and J. J. Oppenheim.** 1989. Macrophage-derived mediators: interleukin 1, tumor necrosis factor, interleukin 6, interferon and related cytokines, p. 639–661. *In* W. E. Paul (ed.), *Fundamental Immunology,* 2nd ed. Raven Press, New York.

20. **Fan, H.** 1990. Influences of the long terminal repeats on retrovirus pathogenicity. *Semin. Virol.* **1**:165–174.

21. **Farrar, J. J., W. R. Benjamin, M. L. Hilfiker, M. Howard, W. L. Farrar, and J. Fuller-Farrar.** 1982. The biochemistry, biology and role of interleukin 2 in the induction of cytotoxic T cells and antibody-forming B cell responses. *Immunol. Rev.* **63**:129–166.

22. **Fauci, A. S.** 1988. The human immunodeficiency virus: infectivity and mechanisms of pathogenesis. *Science* **239**:617–622.

23. **Ghosh, S., A. M. Gifford, L. R. Riviere, P. Tempst, B. P. Nolan, and D. Baltimore.** 1990. Cloning of the p50 DNA binding subunit of NF-$_{kappa}$B: homology to *rel* and *dorsal*. *Cell* **62**:1019–1029.

24. **Gillis, S.** 1989. T-cell-derived lymphokines, p. 621–628. *In* W. E. Paul (ed.), *Fundamental Immunology,* 2nd ed. Raven Press, New York.

25. **Graf, T., and H. Beug.** 1978. Avian leukemia viruses. Interaction with their target cells in vivo and in vitro. *Biochim. Biophys. Acta* **516**:269–299.

26. **Graham, M., J. M. Adams, and S. Corey.** 1985. Murine T lymphomas with retroviral inserts in the chromosome 14 locus for plasmacytoma variant translocations. *Nature* (London) **314**:740–745.

27. **Handa, K., R. Suzuki, H. Matsui, Y. Shimizu, and K. Kumagai.** 1983. Natural killer (NK) cells as a responder to interleukin 2 (IL 2). *J. Immunol.* **130:**988–992.

28. **Hanley, S. M., L. T. Rimsky, M. H. Malim, J. H. Kim, J. Hauber, M. Duc Dodon, S.-Y. Le, J. V. Maizel, B. R. Cullen, and W. C. Greene.** 1989. Comparative analysis of the HTLV-I Rex and HIV-1 Rev *trans*-regulatory proteins and their RNA response elements. *Genes Dev.* **3:**1534–1544.

29. **Harper, M. E., L. M. Marselle, R. C. Gallo, and F. Wong-Staal.** 1986. Detection of lymphocytes expressing human T-lymphotropic virus type III in lymph nodes and peripheral blood from infected individuals by in situ hybridization. *Proc. Natl. Acad. Sci. USA* **83:**772–776.

30. **Hartley, J. W., N. K. Wolford, L. J. Old, and W. P. Rowe.** 1977. A new class of murine leukemia virus associated with development of spontaneous lymphomas. *Proc. Natl. Acad. Sci. USA* **74:**789–792.

31. **Hayward, W. S., B. G. Neel, and S. M. Astrin.** 1981. Activation of a cellular *onc* gene by promoter insertion in ALV-induced lymphoid leukosis. *Nature* (London) **290:**475–480.

32. **Heaphy, S., C. Dingwall, I. Ernberg, M. J. Gait, S. M. Green, J. Karn, A. D. Lowe, M. Singh, and M. A. Skinner.** 1990. HIV-1 regulator of virion expression (Rev) protein binds to an RNA stem-loop structure located within the rev response element region. *Cell* **60:**685–693.

33. **Hidaka, M., J. Inoue, M. Yoshida, and M. Seiki.** 1988. Post transcriptional regulator (*rex*) of HTLV-I initiates expression of viral structural proteins but suppresses expression of regulatory proteins. *EMBO J.* **7:**519–523.

34. **Hirsch, M. S., and J. Curran.** 1990. Human immunodeficiency viruses, p. 1545–1570. *In* B. N. Fields, D. M. Knipe, R. M. Chanock, M. S. Hirsch, J. L. Melnick, T. P. Monath, and B. Roizman (ed.), *Virology,* 2nd ed. Raven Press, New York.

35. **Holland, C. A., C. Y. Thomas, S. K. Chattopadhyay, C. Koehne, and P. V. O'Donnell.** 1989. Influence of enhancer sequences on thymotropism and leukemogenicity of mink cell focus-forming viruses. *J. Virol.* **63:**1284–1292.

36. **Hsu, T. W., J. L. Sabran, G. E. Mark, R. V. Guntaka, and J. M. Taylor.** 1978. Analysis of unintegrated avian RNA tumor virus double-stranded DNA intermediates. *J. Virol.* **28:**810–818.

37. **Hughes, S. E., P. R. Shank, D. H. Spector, H.-J. Kung, J. M. Bishop, H. E. Varmus, P. K. Vogt, and M. L. Breitman.** 1978. Proviruses of avian sarcoma virus are terminally redundant, co-extensive with unintegrated linear DNA and integrated at many sites. *Cell* **15:**1397–1410.

38. **Jacob, R. J., L. S. Morse, and B. Roizman.** 1979. Anatomy of herpes simplex virus DNA. XII. Accumulation of head-to-tail concatemers in nuclei of infected cells and their role in the generation of the four isomeric arrangements of viral DNA. *J. Virol.* **29:**448–457.

39. **Kestler, H., T. Kodama, D. Ringler, M. Marthas, N. Pederson, A. Lackner, D. Regier, P. Sehgal, M. Daniel, N. King, and R. Desrosiers.** 1990. Induction of AIDS in rhesus monkeys by molecularly cloned simian immunodeficiency virus. *Science* **248:**1109–1112.

40. **Kieff, E., and D. Leibowitz.** 1985. Epstein-Barr virus and its replication, p. 1889–1920. *In* B. N. Fields and D. M. Knipe (ed.), *Virology.* Raven Press, New York.

41. **Kiernan, M., V. Blank, F. Logeat, J. Vandekerckhove, F. Lottspeich, O. le Ball, M. B. Urban, P. Kourilsky, P. A. Baeuerle, and A. Israel.** 1990. The DNA binding subunit of NF$_{-kappa}$B is identical to factor KBF1 and homologous to the *rel* oncogene product. *Cell* **62:**1007–1018.

42. **Klein, G., B. C. Giovanella, T. Lindahl, P. J. Fialkow, S. Singh, and J. S. Stehlin.** 1974. Direct evidence for the presence of Epstein-Barr virus DNA and nuclear antigen in malignant epithelial cells from patients with poorly differential carcinoma of the nasopharynx. *Proc. Natl. Acad. Sci. USA* **71:**4737–4741.

43. **Klein, G., and E. Klein.** 1985. Evolution of tumours and the impact of molecular oncology. *Nature* (London) **315:**190–195.

44. **Larn, K., and D. H. Crawford.** Unpublished observations.

45. **Leis, J., D. Baltimore, J. M. Bishop, J. Coffin, E. Fleissner, S. P. Goff, S. Oroszlan, H. Robinson, A. M. Skalka, and H. M. Temin.** 1988. Standardized and simplified nomenclature for proteins common to all retroviruses. *J. Virol.* **62:**1808–1809.

46. **Li, Y., C. A. Holland, J. W. Hartley, and N. Hopkins.** 1984. Viral integration near *c-myc* in 10 to 20% of MCF 247-induced AKIR lymphomas. *Proc. Natl. Acad. Sci. USA* **81:**6808–6811.

47. **Li, Y., E. Golemis, J. W. Hartley, and N. Hopkins.** 1987. Disease specificity of nondefective Friend and Moloney murine leukemia viruses is controlled by a small number of nucleotides. *J. Virol.* **61:**693–700.

48. **Lung, M. L., G. Hering, J. W. Hartley, W. P. Rowe, and N. Hopkins.** 1980. Analysis of the genomes of mink cell focus-inducing sarcoma type C viruses: a progress report. *Cold Spring Harbor Symp. Quant. Biol.* **44:**1269–1274.

49. **Malim, M. H., L. S. Tiley, D. F. McCarn, J. R. Rusche, J. Hauber, and B. R. Cullen.** 1990. HIV-1 structural gene expression requires binding of the Rev *trans*-activator to its RNA target sequence. *Cell* **60:**675–683.

50. **Merrill, J. E., Y. Koyanagi, and I. S. Y. Chen.** 1989. Interleukin 1 and tumor necrosis factor alpha can be induced from mononuclear phagocytes by HIV-1 binding to the CD4 receptor. *J. Virol.* **63:**4404–4408.

51. **Miller, G.** 1985. Epstein-Barr virus: biology, pathogenesis, and medical aspects, p. 1921–1958. *In* B. N. Fields and D. M. Knipe (ed.), *Virology.* Raven Press, New York.

52. **Nusse, R., and H. E. Varmus.** 1982. Many tumors induced by the mouse mammary tumor virus contain a provirus integrated in the same region of the host genome. *Cell* **31:**99–109.

53. **Nyborg, J. K., W. S. Dynan, I. S. Y. Chen, and W. Wachsman.** 1988. Binding of host-cell factors to DNA sequences in the long terminal repeat of human T-cell leukemia virus type I: implications for viral gene expression. *Proc. Natl. Acad. Sci. USA* **85:**1457–1461.

54. **Overbaugh, J., P. R. Donahue, S. L. Quackenbush, E. A. Hoover, and J. I. Mullins.** 1988. Molecular cloning of a feline leukemia virus that induces fatal immunodeficiency disease in cats. *Science* **239:**906–910.

55. **Parker, D. C.** 1982. Separable helper factors support B cell proliferation and maturation to Ig secretion. *J. Immunol.* **129:**469–474.

56. **Paskalis, H., B. K. Felber, and G. N. Pavlakis.** 1986. Cis-acting sequences responsible for the transcriptional activation of human T-cell leukemia virus type I constitute a conditional enhancer. *Proc. Natl. Acad. Sci. USA* **83:**6558–6562.

57. **Payne, G. S., J. M. Bishop, and H. E. Varmus.** 1982. Multiple arrangements of viral DNA and an activated host oncogene in bursal lymphomas. *Nature* (London) **295:**209–214.

58. **Pedersen, N. C., E. W. Ho, M. L. Brown, and J. K. Yamamoto.** 1987. Isolation of a T-lymphotropic virus from domestic cats with an immunodeficiency-like syndrome. *Science* **235:**790–793.

59. **Peterlin, B. M., P. A. Luciw, P. J. Barr, and M. D. Walker.** 1986. Elevated levels of mRNA can account for the *trans*-activation of human immunodeficiency virus (HIV). *Proc. Natl. Acad. Sci. USA* **83:**9734–9738.

60. **Pomerantz, R. J., D. Trono, M. B. Feinberg, and D. Baltimore.** 1990. Cells nonproductively infected with HIV-1 exhibit an aberrant pattern of viral RNA expression: a molecular model for latency. *Cell* **61:**1271–1276.

61. **Ramakrishnan, L., and N. Rosenberg.** 1989. *abl* genes. *Biochim. Biophys. Acta* **989:**209–224.

62. **Reinherz, E. L., P. C. Kung, G. Goldstein, and S. F. Schlossman.** 1979. Separation of functional subsets of human T cells by a monoclonal antibody. *Proc. Natl. Acad. Sci. USA* **76:**4061–4065.

63. **Roitt, I. M., J. Brostoff, and D. K. Male.** 1985. *Immunology.* Gower Medical Publications, London.

64. **Rosen, C. A., J. G. Sodroski, K. Campbell, and W. A. Haseltine.** 1985. The location of *cis*-acting regulatory sequences in the human T cell lymphotrophic virus type III (HTLV-III/LAV) long terminal repeat. *Cell* **41**:813–823.
65. **Rosenberg, N., and O. N. Witte.** 1988. The Abelson (*abl*) oncogene. *Adv. Vir. Res.* **35**:39–81.
66. **Rosenblatt, J. D., A. J. Cann, D. J. Slamon, E. S. Smalberg, N. P. Shah, J. Fujii, W. Wachsman, and I. S. Y. Chen.** 1988. HTLV-II *trans*-activation is regulated by two overlapping nonstructural genes. *Science* **240**:916–919.
67. **Sattentau, Q. J., and R. A. Weiss.** 1988. The CD4 antigen: physiological ligand and HIV receptor. *Cell* **52**:631–633.
68. **Seiki, M., R. Eddy, T. B. Shows, and M. Yoshida.** 1984. Nonspecific integration of the HTLV provirus genome into adult T-cell leukaemia cells. *Nature* (London) **309**:640–642.
69. **Seiki, M., S. Hattori, V. Hirayama, and M. Yoshida.** 1983. Human adult T-cell leukemia virus: complete nucleotide sequence of the provirus genome integrated in leukemia cell DNA. *Proc. Natl. Acad. Sci. USA* **80**:3618–3622.
70. **Selten, B., H. T. Cuypers, M. Ziljstra, C. Blief, and A. Berns.** 1984. Involvement of *c-myc* in M-MuLV-induced T-cell lymphomas of mice: frequency of activation. *EMBO J.* **13**:3215–3222.
71. **Sodroski, J. G., C. A. Rosen, and W. A. Haseltine.** 1984. *Trans*-acting transcriptional activation of the long terminal repeat of human T lymphotropic viruses in infected cells. *Science* **225**:381–385.
72. **Sodroski, J. G., C. A. Rosen, and W. A. Haseltine.** 1985. *Trans*-acting transcriptional regulation of human T-cell leukemia virus type III long terminal repeat. *Science* **227**:171–173.
73. **Stephens, R. M., N. R. Rice, R. R. Hiebsch, H. R. Bose, Jr., and R. V. Gilden.** 1983. Nucleotide sequence of *v-rel:* the oncogene of reticuloendotheliosis virus. *Proc. Natl. Acad. Sci. USA* **80**:6229–6233.
74. **Stinski, M. F.** 1985. Cytomegalovirus and its replication, p. 1959–1980. *In* B. N. Fields and D. M. Knipe (ed.), *Virology.* Raven Press, New York.
75. **Stoye, J. P., and J. M. Coffin.** 1987. The four classes of endogenous murine leukemia virus: structural relationships and potential for recombination. *J. Virol.* **61**:2659–2669.
76. **Teich, N., J. Wyke, T. Mak, A. Berstein, and W. Hardy.** 1982. Pathogenesis of retrovirus-induced disease, p. 785–998. *In* R. Weiss, N. Teich, H. Varmus, and J. Coffin (ed.), *RNA Tumor Viruses. Molecular Biology of Tumor Viruses,* 2nd ed. Cold Spring Harbor Laboratory, Cold Spring Harbor, N.Y.
77. **Thomas, C. Y., and J. M. Coffin.** 1982. Genetic alterations of RNA leukemia viruses associated with the development of spontaneous thymic leukemia in AKR/J mice. *J. Virol.* **43**:416–426.
78. **Tsichlis, P. N., M. A. Lohse, C. Szpirer, J. Szpirer, and G. Levan.** 1985. Cellular DNA regions involved in the induction of rat thymic lymphomas (*Mlvi-1, Mlvi-2, Mlvi-3,* and *c-myc*) represent independent loci as determined by their chromosomal map location in the rat. *J. Virol.* **56**:938–942.
79. **Varmus, H. E., and R. S. Swanstrom.** 1982. Replication of retroviruses, p. 369–512. *In* R. Weiss, N. Teich, H. Varmus, and J. Coffin (ed.), *RNA Tumor Viruses. Molecular Biology of Tumor Viruses,* 2nd ed. Cold Spring Harbor Laboratory, Cold Spring Harbor, N.Y.
80. **Vijaya, S., D. L. Steffen, and H. L. Robinson.** 1987. *Dsi-1,* a region with frequent insertions in Moloney murine leukemia virus-induced rat thymomas. *J. Virol.* **61**:1164–1170.
81. **Watson, J., and D. Mochizuki.** 1981. Interleukin 2: a class of T cell growth factors. *Immunol. Rev.* **51**:257–278.
82. **Wong-Staal, F.** 1990. Human immunodeficiency viruses and their replication, p. 1529–1544. *In* B. N. Fields, D. M. Knipe, R. M. Chanock, M. S. Hirsch, J. L. Melnick, T. P. Monath, and B. Roizman (ed.), *Virology,* 2nd ed. Raven Press, New York.

83. **Wu, F. K., J. A. Garcia, D. Harrich, and R. B. Gaynor.** 1988. Purification of the human immunodeficiency virus type 1 enhancer and TAR binding proteins EBP-1 and UBP-1. *EMBO J.* **7:**2117–2129.

84. **Zack, J. A., S. J. Arrigo, S. R. Weitsman, A. S. Go, A. Haislip, and I. S. Y. Chen.** 1990. HIV-1 intry into quiescent primary lymphocytes: molecular analysis reveals a labile, latent viral structure. *Cell* **61:**213–222.

85. **Zapp, M. L., and M. R. Green.** 1989. Sequence specific RNA binding by the HIV-1 Rev protein. *Nature* (London) **342:**714–717.

86. **Zeng, Y., L. G. Zhang, Y. C. Wu, Y. S. Huang, N. Q. Huang, J. Y. Li, Y. B. Wang, M. K. Jiang, Z. Fang, and N. N. Ming.** 1985. Prospective studies on nasopharyngeal carcinoma in EBV IgA/VCA gamma positive persons in Wuzhou City, China. *Int. J. Cancer* **36:**545–547.

87. **Zubler, R. H., J. W. Lowenthal, F. Erard, N. Hashimoto, R. Devos, and H. R. MacDonald.** 1984. Activated B cells express receptors for, and proliferate in response to, pure interleukin 2. *J. Exp. Med.* **160:**1170–1183.

Part I

AUTOIMMUNITY

Viruses That Affect the Immune System
Edited by Hung Y. Fan et al.
© 1991 American Society for Microbiology, Washington, DC 20005

Chapter 2

Virus-Induced Autoimmunity

Michael B. A. Oldstone

Viruses both activate and magnify autoimmunity, as established by at least three findings. First, in humans autoimmune responses are made de novo, or those already present are enhanced, concomitant with infection by a wide variety of DNA and RNA viruses. Second, both acute and persistent virus infections in experimental animals can induce, accelerate, or enhance autoimmune responses and cause autoimmune disease (13). For example, the volume of anti-DNA antibodies normally present in (NZB × NZW)F$_1$ mice is markedly enhanced by persistent infection with polyomavirus, a DNA virus, or lymphocytic choriomeningitis virus, an RNA virus (22). Further, NZW mice, which normally do not develop these autoimmune responses, do so upon infection with polyomavirus or lymphocytic choriomeningitis virus (9, 22). A number of viruses, including retroviruses, are now known to perform similarly. Third, evaluation of molecular mimicry (see below) in human autoimmune disorders has uncovered a number of potential etiologic agents and mechanisms of autoimmune disease (14).

Viruses induce autoimmune responses by several means. For example, certain viruses have a mitogenic effect on blood lymphocytes and act as polyclonal activators. Viruses also direct the release of cytokines, which then modulate immune responses by acting directly as growth or differentiation factors or by regulating the expression of class I and class II major histocompatibility molecules. Viruses can also replicate selectively in particular lymphocyte subsets. By their presence, activation, or replication they can cause immunosuppression or immunoenhancement. Moreover, some viruses and

Michael B. A. Oldstone • Division of Virology, Department of Neuropharmacology, Research Institute of Scripps Clinic, 10666 N. Torrey Pines Road, La Jolla, California 92037.

other microbes contain chemical structural components that mimic normal host "self" proteins. An effector immune response, of either the B-cell (antibodies) or T-cell (cytotoxic T cell) variety, directed against the microbe might then also cross-react with self protein, thereby evoking autoimmunity.

Molecular mimicry is defined as similar structures shared by molecules from dissimilar genes or by their protein products. Either the molecules' linear amino acid sequences or their conformational fit may be shared, even though their origins are as separate as, for example, a virus and a normal host self determinant. Because guanine-cysteine (GC) sequences and introns designed to be spliced away may provide, respectively, false hybridization signals and nonsense homologies, molecular mimicry is best analyzed at the protein level. Such homologies between proteins have been detected either by use of immunologic reactants, humoral or cellular, that cross-react with two presumably unrelated protein structures, or by matching proteins described in computer storage banks. Regardless of the methods used for identification, it is now abundantly clear that molecular mimicry is common between proteins encoded by numerous DNA and RNA viruses and host self proteins. Such events are relevant not only to autoimmunity but also as a likely mechanism by which viral proteins are processed inside cells (2).

Molecular mimicry occurs frequently and with a wide variety of DNA and RNA viruses. Over 800 monoclonal antibodies raised against viral polypeptides were tested for cross-reactivity with host proteins expressed in a large panel of normal tissues (Table 1). Nearly 5% of monoclonal antibodies reactive with 14 different DNA and RNA viruses—including herpesvirus, vaccinia virus, myxoviruses, paramyxoviruses, arenaviruses, flaviviruses, alphaviruses, rhabdoviruses, coronaviruses, and human retroviruses—cross-reacted with host cell determinants expressed on uninfected tissues. Some of these monoclonal antiviral antibodies reacted with constituents of more than one organ (20). Hence, molecular mimicry is common and not restricted to any specific class or group of virus. Such extensive cross-reactivity raises provocative questions about the etiology and pathogenesis of numerous autoimmune diseases (Table 1).

Molecular mimicry is clearly capable of eliciting autoimmune diseases (6, 12). Myelin basic protein was initially chosen as the host component to examine for study of this issue (6) because its entire amino acid sequence is known and its encephalitogenic site of 8 to 10 amino acids (aa) has been mapped in several animal species. Computer-assisted analysis of the Dayhoff files showed that several viral proteins had significant homology with the encephalitogenic site of myelin basic protein. Included were similarities and/ or fits between myelin basic protein and the nucleoprotein and hemagglutinin of influenza virus, coat protein of polyomavirus, core protein of the adenovirus, polyprotein of poliomyelitis virus, EC-LF2 protein of Epstein-Barr

Table 1. Molecular Mimicry between Antiviral Monoclonal Antibodies and
Host Self Antigens[a]

Monoclonal antibodies reactive with virus:	No. tested	No. reactive with uninfected (self) tissue
DNA viruses		
Herpes simplex virus type 1	21	2
Cytomegalovirus		
Human	24	1
Mouse	14	0
Vaccinia virus	16	1
RNA viruses		
Measles virus	39	5
Rabies virus	80	2
Vesticular stomatitis virus	37	2
Lymphocytic choriomeningitis virus	174	3
Coxsackievirus type B	66	1
Theiler's virus	64	9
Japanese encephalitis virus	34	6
Dengue virus	132	0
Human immunodeficiency virus	128	8
Total	829	40 (4.8%)

[a]Initial study on over 600 monoclonal antibodies was done at the National Institutes of Health and reported by Srinivasappa et al. (see reference 20 for details). These monoclonal antibodies were generated at the National Institutes of Health, Walter Reed Army Institute of Research, Wistar Institute, and the Research Institute of Scripps Clinic. The additional monoclonals reported are primarily to cytomegalovirus, human immunodeficiency virus, and Theiler's virus and were generated and tested at the Research Institute of Scripps Clinic. Human immunodeficiency virus monoclonal antibodies were tested primarily against mammalian central nervous system tissue. A number of these cross-reactivities offered intriguing hints of viral pathogenesis. For example, the coxsackievirus type B monoclonal antibody not only neutralized the virus but also cross-reacted with mammalian myocardial tissue. A monoclonal antibody to Theiler's virus VP-1 neutralized the virus, stained a major determinant on the surface of oligodendrocytes (galactocerebroside), and caused demyelination when inoculated into the myelin sheath. One of the monoclonal antibodies to measles virus stained a novel determinant on the surface of a subset of T lymphocytes, and measles virus infection is associated with immunosuppression. A number of monoclonal antibodies to human immunodeficiency virus gp41 immunodominant domain stained a novel 42- to 43-kDa protein on a subset of astrocytes. Since astrocytes are believed to be involved in maintenance of the appropriate neuronal milieu and act as a component of the blood-brain barrier, their dysfunction might lead to the reversible pharmacologic dementia seen in AIDS.

virus, hepatitis B virus polymerase (HBVP), and others. However, the best fit occurred between the myelin basic protein encephalitogenic site in the rabbit and HBVP (6; Fig. 1).

Interestingly, products of immune responses, both humoral and cellular, generated in rabbits inoculated with the octamer or decamer viral peptide reacted with whole myelin basic protein. Further, inoculation of the HBVP peptide into rabbits caused perivascular infiltration localized to the central nervous system (6; Fig. 1), reminiscent of the disease induced by inoculation of either whole myelin basic protein or the encephalitogenic site of myelin

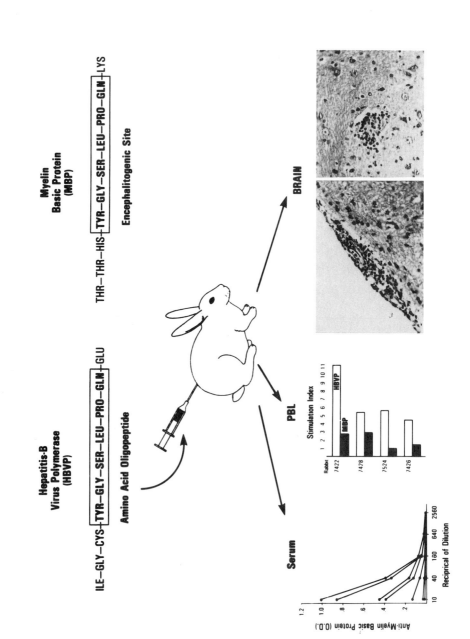

basic protein. Afterward, it was shown that coxsackievirus type B shared amino acid sequences with myosin (12). During coxsackievirus infection in A strain mice, antimyosin antibodies were generated. When such virus-free antibodies were adoptively transferred, they caused myocarditis (12; Fig. 2). Both these experiments were seminal, since they conclusively showed that molecular mimicry causes autoimmune responses and autoimmune disease.

The most likely explanation for how molecular mimicry causes disease is that an immune response against the determinant shared by host and virus takes the form of a tissue-specific attack, presumably capable of destroying cells and eventually the tissue. The probable mechanism is generation by the pathogen of cytotoxic cross-reactive effector lymphocytes or antibodies that recognize specific determinants of self proteins located on target cells. Interestingly, the induction of cross-reactivity would not require a replicating agent, and the immunologically mediated injury could occur after removal of the pathogen, a hit-and-run event. Clearly, the virus infection that initiates an autoimmune phenomenon need not be present at the time overt disease develops. A likely scenario would be that the virus responsible for inducing a cross-reacting immune response is cleared initially, but that components of that response continue to assault host elements. The cycle continues as the autoimmune response itself leads to tissue injury that, in turn, releases more self antigen, thereby inducing more antibodies, and so on. Such a sequence might account for the viral encephalopathies occurring in humans after measles, mumps, vaccinia, or herpes zoster virus infections; in these postinfectious diseases, recovery of the inducing agent has been rare.

This theory is reinforced by studies showing that, after any of several acute viral infections, mononuclear cells from the peripheral blood or cerebrospinal fluid proliferate in response to host antigens, one of which is myelin basic protein. Interestingly, several populations of lymphocytes harvested from central nervous system fluids of humans with encephalitis proliferated clonally in response to the infecting virus as well as to antigens of the nervous system (reviewed in reference 7).

Viruses or microbes also play by other game plans. For example, a virus with the capacity to persist in its host may continuously or cyclically express its antigens. Although expression of a viral genome may be restricted so that

---

Figure 1. Initial evidence that molecular mimicry can cause autoimmune responses and autoimmune disease (see reference 6). When New Zealand rabbits were inoculated with the 10-aa peptide from HBVP, they generated specific T (proliferation)- and B (antibody)-lymphocyte responses. Most significant, 4 of the 11 inoculated rabbits (40%) developed histopathologic criteria for lesions of allergic encephalomyelitis. In contrast, studies with over 10 different peptides in more than 30 rabbits failed to elicit allergic encephalomyelitis. O.D., optical density; PBL, peripheral blood lymphocytes.

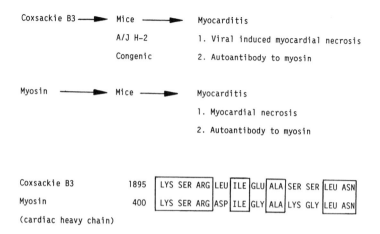

Figure 2. Additional evidence that molecular mimicry can cause autoimmune mediated disease. Schematic of data published by Neu et al. (see reference 12) showing molecular mimicry between coxsackievirus type B and cardiac myosin. This work followed the early demonstration by Srinivasappa and colleagues (see reference 20) that a monoclonal antibody capable of neutralizing coxsackievirus type B also stained myocardial tissue from several mammalian species.

no infectious virus replicates, production of a viral determinant in common with that of the host might continue. This could initiate an immune response or autoimmunity, either one leading to cyclic, chronic, or progressive disease.

In any case, molecular mimicry would occur only when the virus and host determinants are sufficiently similar to induce a cross-reactive response yet different enough to break B- or T-cell immunologic tolerance.

Clinical evidence for the hypothesis that molecular mimicry causes autoimmune disease in humans is difficult to come by, although a number of interesting possibilities are worth exploring (Fig. 3). Some evidence now links immune reactants with specific diseases, for instance, antibodies to acetylcholine α-chain aa 160–167 with myasthenia gravis and herpes simplex virus gpD aa 286–293; α-gliadin and adenovirus type 12 with celiac disease; HLA-B27 and *Klebsiella pneumoniae* nitrogenase, *Shigella flexneri, Yersinia pseudotuberculosis,* and *Yersinia enterocolitica* with nonrheumatoid arthritides like ankylosing spondylitis or Reiter's syndrome (1, 5, 8, 15, 18, 19, 23). The approach taken and the results with one of these studies, the association between acetylcholine receptor (AChR)-herpes simplex virus and myasthenia gravis, are summarized below.

Patients with the autoimmune disease myasthenia gravis characteristically have detectable antibodies against AChR. The nicotinic AChR is composed of multiple subunits responsible for gating ion flow across membranes in response to binding of the neurotransmitter acetylcholine. The actions of

HuAChR   160:   PRO GLU SER ASP GLN PRO ASP LEU
HSV GP-D   286:   PRO ASN ALA THR GLN PRO GLU LEU

α-GLIADIN   206:   LEU ARG ARG GLY MET PHE ARG PRO SER GLN CYS ASN
ADENO 12 E1b 384:   LEU GLN ARG GLY SER PHE ARG PRO SER GLN GLN ASN

HLA-B27   70:   LYS ALA GLN THR ASP ARG GLU ASP LEU
KLEB PN   186:   SER ARG GLN THR ASP ARG GLU ASP GLU

INSULIN r   66:   VAL TYR GLY LEU GLU SER LEU LYS ASP LEU
PAPILLOMA E2 76:   VAL LEU HIS LEU GLU SER LEU LYS ASP SER

HLA DR   50:   VAL THR GLU LEU GLY ARG PRO ASP ALA GLU
HCMV IE-2   79:   PRO ASP PRO LEU GLY ARG PRO ASP GLU ASP

COAGFACT XI   269:   ILE LYS LYS SER LYS ALA LEU
DENGUE   68:   ILE LYS LYS SER LYS ALA ILE

MYOSIN   138:   TYR GLU ALA PHE VAL LYS HIS ILE MET SER VAL
COX B3   2152: TYR GLU ALA PHE ILE ARG LYS ILE ARG SER VAL

BRAIN PROTEIN 156:   ASP SER THR LYS ASN ARG LYS THR ASP
HIV POL   222:   ASP SER THR LYS TRP ARG LYS VAL ASP

Figure 3. Sequence similarities between microbial proteins and human host self proteins. Examples of interesting shared sequences under active evaluation for pathogenic role in selected diseases. Significant immunologic cross-reactivity has now been observed for the majority of these sequence pairs. Disease association has been suggested for the first four pairs by investigation of specimens from patients with myasthenia gravis, celiac disease, ankylosing spondylitis, Reiter's syndrome, and acanthosis nigricans. The latter four pairs are of interest because of associations of the ASP in aa 57 of HLA DR for patients with diabetes (21) and recent preliminary work showing human cytomegalovirus (HCMV) sequences persisting in the islets of Langerhans of a subset of diabetic patients (10); the association of dengue virus infection, hemorrhagic fever, and coagulation proteins; the findings of coxsackievirus (COX)-induced myocarditis (see Fig. 2 and reference 12); and the reversible (with azidothymidine treatment) dementia complex in patients with AIDS. The novel brain protein was isolated and sequenced in J. Gregor Sutcliffe's laboratory at the Research Institute of Scripps Clinic. HSV GP-D, herpes simplex virus glycoprotein D; ADENO 12, adenovirus type 12; KLEB PN, *Klebsiella pneumoniae;* COAGFACT XI, coagglutination factor XI; HIV POL, human immunodeficiency virus polymerase.

AChR antibodies, either experimentally induced or spontaneous, lead to the numerical reduction of available AChR and prevention of the neuromuscular junction's ability to transmit signals from nerve fibers to muscle fibers. Medically, the outcome is myasthenia gravis. For molecular dissection of the physiologically important binding sites, synthetic peptides representing unique regions of the α-chain of AChR have been used in association with α-bungarotoxin to compete with the binding of antibody specific for AChR. These interactions also map accessible sites on AChR that bind antibodies. We (18) found immunochemically significant cross-reactivity with the human AChR α-chain aa 160–167 sequences and herpes simplex virus glycoprotein D residues 286–293 (Fig. 3 and 4). For example, antibodies to the herpes simplex virus peptide aa 286–293 reacted with both the corresponding AChR peptide and the AChR native protein. Interestingly, a different herpes simplex

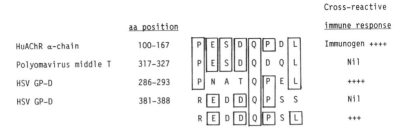

Figure 4. Importance of leucine in position 293 for binding of herpes simplex virus glycoprotein D (HSV GP-D) peptide aa 286–293 or failure of binding of herpes simplex virus glycoprotein D peptide aa 381–388 to antibody to human AChR (HuAChR) α-chain. Individually synthesized peptides were reactive with antibody to human AChR α-chain aa 160–167. Binding data were confirmed by competition assay with free peptide (see reference 4).

virus glycoprotein, aa 381–389, also shared several similar amino acid sequences with the AChR (Fig. 4). However, immunochemical analysis failed to document functional cross-reactivity. Using single-amino-acid substitutions revealed that a single amino acid, leucine, in position 293 of herpes simplex virus glycoprotein D is critical in the cross-reactivity with AChR 160–167 (4). Affinity purification of antibodies from patients with myasthenia gravis, using the human AChR α-chain 157–170 peptide immobilized on thiopropyl-Sepharose, yielded immunoglobulin G antibodies that bound to the native AChR and inhibited the binding of α-bungarotoxin to its specific binding site on the receptor. Thus, the human AChR α-chain 160–167 peptide specifically cross-reacted with a shared homologous domain on herpes simplex virus glycoprotein D, residues 286–293, as shown by binding and inhibition studies (18). Further, antibodies elicited in myasthenic patients bound to AChR 160–167, herpes simplex virus glycoprotein D 286–293 protein, and native herpes simplex virus protein. Finally, these cross-reactive antibodies caused a biologic effect on AChR. The immunologic cross-reactivity of this self epitope with herpes simplex virus suggests that this virus may be associated with some cases of myasthenia.

The probability that a random six-amino-acid sequence will be identical in two dissimilar proteins is 1 in 2×10^6, assuming that all amino acids are represented equally. Nevertheless, the finding of sequence homology is, by itself, insufficient evidence of biologically meaningful mimicry. The study with AChR illustrates this point. Despite a high degree of similarity between portions of the AChR α-chain (PESDQPDL) and the polyomavirus middle T antigen (PESDQDQL), no cross-reacting antibodies formed (Fig. 4). However, in another setting, a more distant similarity between the AChR sequence and the herpes simplex virus glycoprotein D (PNATQPEL) induced strong immunologic cross-reactivity (4, 18; Fig. 4).

Just as homologies and immunologic cross-reactivities have been found between host and microbial proteins, as listed in Fig. 3, additional similarities will surely emerge as more genes and proteins are analyzed (3, 11, 14, 16, 17). Some of these examples may well account for diseases in terms of an autoimmune response provoked by molecular mimicry. Probing that association will be difficult, but an exciting challenge for the future.

Acknowledgments. This is publication number 6234-NP from the Department of Neuropharmacology, Scripps Clinic and Research Foundation, La Jolla, Calif. This work was supported in part by Public Health Service grants AI-07007, AG-04342, and NS-12428. I acknowledge the contributions made by T. Dyrberg and P. Schwimmbeck during their postdoctoral fellowships and by Robert Fujinami, a long-time colleague.

REFERENCES

1. **Chen, H.-J., D. H. Kono, A. Young, M. S. Park, M. B. A. Oldstone, and D. T. Y. Yu.** 1987. A *Yersinia pseudotuberculosis* protein which reacts with HLA B27. *J. Immunol.* **139:**3003–3011.
2. **Dales, S., R. S. Fujinami, and M. B. A. Oldstone.** 1983. Serologic relatedness between Thy1.2 and actin revealed by monoclonal antibody. *J. Immunol.* **131:**1332–1338.
3. **Dyrberg, T., and M. B. A. Oldstone.** 1986. Peptides as probes to study molecular mimicry and virus induced autoimmunity. *Curr. Top. Microbiol. Immunol.* **130:**25–39.
4. **Dyrberg, T., J. S. Petersen, and M. B. A. Oldstone.** 1990. Immunological cross-reactivity between mimicking epitopes on a virus protein and a human autoantigen depends on a single amino acid residue. *Clin. Immunol. Immunopathol.* **54:**290–297.
5. **Ebringer, A.** 1983. The cross-tolerance hypothesis, HLA B27 and ankylosing spondylitis. *Br. J. Rheumatol.* **22**(Suppl. 2):53–66.
6. **Fujinami, R. S., and M. B. A. Oldstone.** 1985. Amino acid homology and immune responses between the encephalitogenic site of myelin basic protein and virus: a mechanism for autoimmunity. *Science* **230:**1043–1045.
7. **Johnson, R. T., and D. E. Griffin.** 1986. Virus-induced autoimmune demyelinating disease of the central nervous system, p. 203–209. *In* A. L. Notkins and M. B. A. Oldstone (ed.), *Concepts in Viral Pathogenesis II.* Springer-Verlag, New York.
8. **Kagnoff, M. F.** 1984. Possible role for a human adenovirus in the pathogenesis of celiac disease. *J. Exp. Med.* **160:**1544–1557.
9. **Lampert, P. W., and M. B. A. Oldstone.** 1973. Host IgG and C3 deposits in the choroid plexus during spontaneous immune complex disease. *Science* **180:**408–410.
10. **Löhr, J. M., and M. B. A. Oldstone.** 1990. Detection of cytomegalovirus nucleic acid sequences in the pancreas' of patients with type II diabetes. *Lancet* **336:**644–648.
11. **Maul, G. G., S. A. Jimenez, E. Riggs, et al.** 1989. Determination of an epitope of the diffuse systemic sclerosis marker antigen DNA topoisomerase I: sequence similarity with retroviral p30^{gag} protein suggests a possible cause for autoimmunity in systemic sclerosis. *Proc. Natl. Acad. Sci. USA* **86:**8492–8496.
12. **Neu, N., N. R. Rose, K. W. Beisel, A. Herskowitz, G. Gurri-Glass, and S. Craig.** 1987. Cardiac myosin induces myocarditis in genetically predisposed mice. *J. Immunol.* **139:**3630–3636.
13. **Oldstone, M. B. A.** 1972. Virus induced autoimmune disease: viruses in the production and prevention of autoimmune disease, p. 469–475. *In* S. R. Day and R. A. Good (ed.), *Membranes and Viruses in Immunopathology: Proceedings.* Academic Press, Inc., New York.

14. **Oldstone, M. B. A.** 1987. Molecular mimicry and autoimmune disease. *Cell* **50**:819–820.
15. **Prendergast, J. K., J. S. Sullivan, A. Geczy, L. I. Upfold, J. P. Edmonds, H. V. Bashir, and E. Levy-Reiss.** 1983. Possible role of enteric organisms in the pathogenesis of ankylosing spondylitis and other seronegative arthropathies. *Infect. Immun.* **41**:935–941.
16. **Query, C. C., and J. D. Keene.** 1987. A human autoimmune protein associated with U1 RNA contains a region of homology that is cross-reactive with retroviral p30^{gag} antigen. *Cell* **51**:211–220.
17. **Rucheton, M., H. Graafland, H. Valles, et al.** 1987. Human autoimmune serum antibodies against gag gene p30 retroviral protein also react with a U1-SnRNP 68K comigrant protein. *Biol. Cell* **60**:71–72.
18. **Schwimmbeck, P. L., T. Dyrberg, D. Drachman, and M. B. A. Oldstone.** 1989. Molecular mimicry and myasthenia gravis: an autoantigenic site of the acetylcholine receptor α-subunit that has biologic activity and reacts immunochemically with herpes simplex virus. *J. Clin. Invest.* **84**:1174–1180.
19. **Schwimmbeck, P. L., D. T. Y. Yu, and M. B. A. Oldstone.** 1987. Autoantibodies to HLA-B27 in the sera of HLA-B27 patients with ankylosing spondylitis and Reiter's syndrome: molecular mimicry with *Klebsiella pneumoniae* as potential mechanism of autoimmune disease. *J. Exp. Med.* **166**:173–181.
20. **Srinivasappa, J., J. Saegusa, B. S. Prabhakar, M. K. Gentry, M. J. Buchmeier, T.J. Wiktor, H. Koprowski, M. B. A. Oldstone, and A. L. Notkins.** 1986. Molecular mimicry: frequency of reactivity of monoclonal antiviral antibodies with normal tissues. *J. Virol.* **57**:397–401.
21. **Todd, J. A., J. I. Bell, and H. O. McDevitt.** 1987. HLA-DQ$_B$ gene contributes to susceptibility and resistance to insulin-dependent diabetes mellitus. *Nature* (London) **329**:599.
22. **Tonietti, G., M. B. A. Oldstone, and F. J. Dixon.** 1970. The effect of induced chronic viral infections on the immunologic diseases of New Zealand mice. *J. Exp. Med.* **132**:89–109.
23. **van Bohemen, C. H. G., F. C. Grumet, and H. C. Zanen.** 1984. Identification of HLA B27 M1 and M2 cross-reactive antigens in *Klebsiella, Shigella,* and *Yersinia. Immunology* **52**:607–610.

Part II

IMMUNODEFICIENCY BY
RETROVIRUSES

Chapter 3

Role of Regulatory Genes in Human Immunodeficiency Virus Replication and Pathogenesis

Flossie Wong-Staal

Human retroviruses have developed a unique strategy to differentially and temporally regulate the expression of different viral proteins. The early proteins are a subset of the regulatory gene products, including the two essential genes for replication, *tat* and *rev* for human immunodeficiency virus (HIV) and *tax* and *rex* for the human T-cell leukemia viruses. The late proteins include those that are incorporated into the mature virion, the core antigens, the enzymes reverse transcriptase, integrase, and endonuclease, and the envelope glycoproteins and, in the case of HIV, also additional accessory gene products. This early-to-late switching is not seen in most retroviruses and is distinct from the early-late switching of DNA viruses, which occurs at the transcriptional level through the use of different promoter elements. In contrast, the mRNA of human retroviruses is expressed from a single primary transcript, and the regulation occurs at a posttranscriptional level. The viral gene that mediates this process is the *rev* gene of HIV or the *rex* gene of human T-cell leukemia virus (Fig. 1). In a single cycle of replication, measured either by a one-step growth experiment (15) or transfection of a non-CD4-containing cell (33), one can see first the appearance of the lower-molecular-weight RNA, which then shifts to include the higher-molecular-weight species. However, in cells transfected with a *rev*-defective virus, this shift does not take place, and only the lower-molecular-weight species accumulate (7, 26). Figure 2 gives a hint of the kind of complexity that one encounters in trying

Flossie Wong-Staal • University of California, San Diego, La Jolla, California 92093.

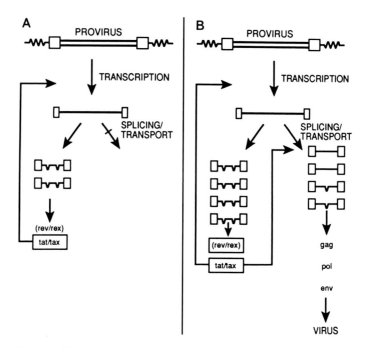

Figure 1. Strategy of human retrovirus gene expression. Viral mRNAs are divided into two classes, the early regulatory mRNAs which can accumulate in the absence of *rev/rex* activity (A) and the late mRNAs which require *rev/rex* (B).

to delineate the species of HIV mRNA expressed in the infected cell. There is quite a bit of redundancy in that different species of mRNA can give rise to the same gene product (27). This is particularly true for some of the regulatory proteins.

   One obvious feature that separates the early mRNAs from the later mRNAs is their extent of splicing. The early mRNAs are multiply spliced, while the late mRNAs are either unspliced or spliced only once. Another feature that stands out is the presence of genetic elements, designated CRS, or *cis*-acting repressive sequences (22), and RRE for *rev*-response element (18), which are found only in the late mRNA and are spliced out in the early mRNA. These sequences confer, respectively, *rev* dependence and *rev* responsiveness.

   The mechanism of Rev protein function is still not ascertained. When my colleagues and I first observed the *rev* mutant phenotype, we proposed that Rev may interfere with the splicing process (7). Recently, a more specific proposal has been made that Rev may disengage RRE-containing viral mRNA from the spliceosomes (1). According to this model, recognition of

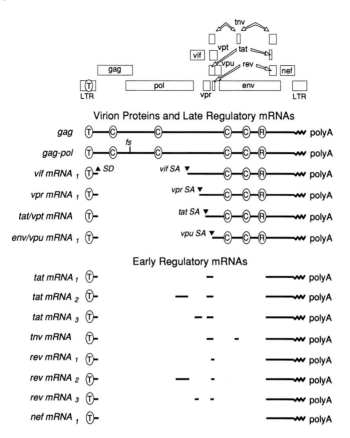

Figure 2. The HIV type 1 genome and mRNA. A schematic diagram of the functional open reading frames and spliced mRNAs is depicted.

individual splice sites (donor or acceptor) in addition to the presence of RRE is a must for Rev responsiveness, and the efficiency of splicing is a determining factor since rapid splicing due to the presence of efficient splice sites precluded Rev responsiveness. There are also data that suggest that *rev* may act at the level of nuclear export, based on the observation that under some circumstances, the effect of Rev is primarily on the cytoplasmic mRNA and that the levels of spliced and unspliced viral RNA remained unchanged in the nucleus in the absence or presence of Rev (21). This apparent disagreement, like the proverbial elephant, may depend on perspective. Splicing and transport may be two processes in a dynamic equilibrium, and Rev, interacting with RRE, may affect the equilibration of these two processes to favor export over splicing. The role of the CRS sequences is still vague; they have been proposed to preferentially trap RNA in the nucleus.

There is now overwhelming evidence that purified Rev protein will bind directly to RRE (3, 4, 14, 19, 34). RRE was originally defined as a 250-nucleotide stretch in the gp41 coding region (18, 23) and is expected to have a high degree of secondary structure (18) (Fig. 3). My colleagues and I have mapped the binding site using restriction fragments of the template DNA and found that the first 90 nucleotides generated by a subfragment (LT-D1) bind Rev as well as the entire 250-nucleotide fragment (LT) (3). Figure 4 shows that both labeled LT and LT-D1 transcripts can be brought down as a complex with Rev by anti-Rev antibodies and, further, that the eluted protein migrated as authentic Rev in both cases. Recognizing that structure, rather than primary sequence, may be important for Rev-RRE interaction, we were surprised to find that LT-D1 would be efficient in binding. When the sequence was analyzed by the algorithm of Zuker (35), it was found that indeed LT-D1 can potentially form a long stem structure which resembles the long stem of RRE. To directly address whether it was this mimicry which is responsible for the binding, we made transcripts from either one strand of

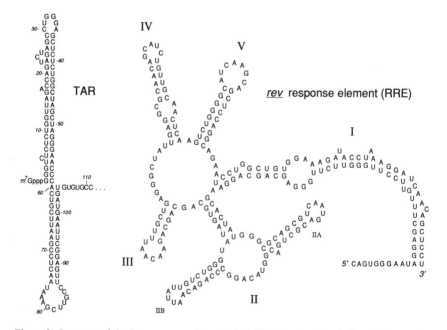

Figure 3. Structure of the Rev response element. A stable stem structure is formed by complementarity of the 5′ and 3′ termini of RRE. This is designated stem I. A series of stem-loop structures intervene and are designated in a clockwise fashion.

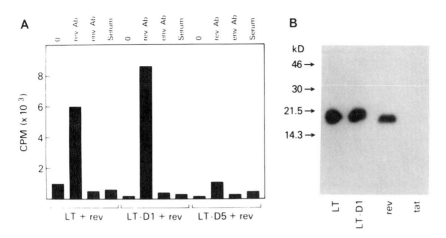

Figure 4. Binding of Rev protein to LT and LT-D1 transcripts. [^{32}P]UTP-labeled RNA transcript of LT and LT-D1, immunoprecipitated with anti-Rev serum, eluted and analyzed on polyacrylamide gels.

the stem or both strands with a few intervening nucleotides to allow looping back of the stem. However, neither of these transcripts could bind Rev (data not shown). When the sequence was reanalyzed using a different program (PC fold), an entirely different structure for LT-D1 was revealed, which now forms a hammerhead structure resembling the hammerhead of LT. Interestingly, all the LT subfragments which bind Rev have a similar predicted structure, while the ones that do not have no corresponding structure. This finding is consistent with those of others which suggest that the hammerhead structure is important for binding and function. However, since the sequence of the LT-D1 hammerhead is quite different from that of the LT hammerhead, the observation supports the idea that the determination at least for binding is largely structural.

Figure 5 shows the predicted amino acid sequence of the Rev protein. Site-directed mutagenesis studies have revealed two important functional domains: a basic domain which serves two functions, for nucleolar localization (2, 8) as well as binding to RRE (20), and a leucine-rich domain which serves as the activating domain since mutations in this region have a negative dominant phenotype (17). By analogy to other transactivators, it is speculated that this domain may interact with a specific cellular factor. However, we and others have failed to identify a cell protein that specifically binds to RRE. We then hypothesized that in order to interact with a cellular factor, Rev must first be brought into proximity with that factor on the RRE, so we started looking for an RRE binding protein. Using nuclear extracts from HeLa

MAGRSGDSDEDLLKAVRLIKFLYQSNPPPN

PEGTRQARRNRRRRWRERQRQIHSISERIL
Nuclear/Nucleolar localization
RRE-RNA binding domain

STYLGRSAEPVPLQLPPLERLTLDCNEDCG
Activator domain

TSGTQGVGSPQILVESPTILESGAKE

Figure 5. Predicted amino acid sequence and functional domains of the Rev protein.

cells, we detected a specific interaction with the LT-D1 transcript using the gel retardation assay (30). This interaction is competed by the homologous RNA, LT and LT-D2, but not by control RNAs like CAT, poly(A), tRNA, or rRNA, nor by the sense strand LT-D1 DNA. Therefore, the protein(s) that interacts with LT-D1 is an RRE RNA binding protein. When we used the LT transcript, we found a similar pattern; in particular, the interaction was completely competed by LT-D1, suggesting that the region of RRE that binds cellular factors is confined to the first 90 nucleotides (data not shown).

UV cross-linking experiments revealed a specific protein of approximately 56 kDa which binds both LT and LT-D1. This protein is competed by the homologous probes, but more importantly, it is the single major band that has not been detected to bind LT-D5, LT-D6, and antisense LT-D1 (Fig. 6). We designated this protein NF_{RRE}. NF_{RRE} seems to be widely distributed among mammalian species, as it was found in every human and rodent nuclear extract we examined. NF_{RRE} also binds with Rev to RRE as a single complex (data not shown). This factor is likely to be important for *rev*-mediated transactivation as well as cellular mRNA processing and transport.

So we have identified at least three players in the Rev transactivation pathway: a viral protein and a cellular protein, both interacting with the same viral RNA target and most likely also with each other. One can design agents to interfere with these interactions. For example, a mutant protein which can still bind to target RNA but is unable to interact with the cell protein can act as a competitive inhibitor of transactivation. Likewise, a complementary (antisense) nucleic acid sequence can disrupt the secondary structure of the target RNA and abolish its activity. Indeed, *rev* mutations in the activator domain, which retains its RNA binding activity, have been shown to have a negative dominant phenotype and can inhibit Rev function and virus rep-

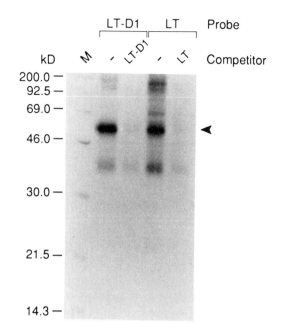

Figure 6. Identification of nuclear factor(s) interacting with RRE. UV cross-linking of radio-labeled LT-D1 or LT RNA to nuclear factors in the presence or absence of unlabeled homologous RNA.

lication (17). Similarly, we have designed antisense oligonucleotides directed at the functionally important regions of RRE and found them to be also inhibitory (32). The translation of these positive results from the laboratory to the clinic will depend on improvement on technology for gene therapy, greater economy in chemical syntheses of some of these compounds, and of course toxicity and pharmacokinetic studies.

tat is the second essential regulatory gene of HIV. There are many parallels between the *tat* and *rev* transactivation pathways. *tat* also contains an activating domain (22) and a nuclear localization/RNA binding domain (13, 25). Like Rev, Tat also recognizes an RNA target sequence, called TAR, which is highly structured (9, 24). Direct binding of Tat to TAR has been shown (5), and the requirement for one or more cellular factors for activity has been implicated (12). Therefore, the antiviral strategies outlined above for Rev would also apply to Tat. In spite of these parallels, the mechanism of *tat* regulation is very different. The primary phenotype of *tat* is to increase gene expression from the virus LTR, and it appears to do so at multiple levels: from increasing the steady-state level of viral mRNA through transcriptional and posttranscriptional mechanisms to increasing the efficiency of translation (see reference 31 for review). At the transcriptional level, *tat* both increases the frequency of initiation and stabilizes the elongation of the growing chain (16). It seems to be a contradiction of terms to have an activator of transcriptional initiation which recognizes RNA, since initiation would have already taken place. It has been postulated that Tat-TAR interaction in the proximity of the promoter can facilitate formation of transcription complexes for subsequent rounds of initiation (28). This mechanism would be completely novel. It is likely that Tat has tapped into an existing cellular pathway for this novel mode of regulation. Again, study of HIV regulation will allow us to gain entry into this untapped area.

Until recently, Tat has been thought of only as a nuclear transactivator which turns on expression of virus in the infected cell. However, recent studies suggest that *tat* may have a much broader impact on HIV pathogenesis. Tat can affect uninfected cells through release and uptake (6, 10). It can activate heterologous genes, including those of a DNA virus (JC virus) which induces a neurological disorder in humans (29). It also functions as an exogenous factor that stimulates cell growth, specifically Kaposi's sarcoma cells, and can be therefore at least a contributory factor for Kaposi's sarcoma pathogenesis (6). Most interestingly, following the lead of these very tantalizing observations, other investigators now find that another viral transactivator protein, namely, Tax of human T-cell leukemia virus type I, also has very similar properties (11). Thus the potential role of these proteins in the overall pathogenesis of the human retroviruses is greatly expanded.

REFERENCES

1. **Chang, D., and P. Sharp.** 1989. Regulation by HIV rev depends upon recognition of splice sites. *Cell* **59:**789–795.
2. **Cullen, B., J. Hauber, K. Campell, J. Sodroski, W. Haseltine, and C. Rosen.** 1988. Subcellular localization of the human immunodeficiency virus *trans*-acting *art* gene product. *J. Virol.* **62:**2498–2501.
3. **Daefler, S., M. Klotman, and F. Wong-Staal.** 1990. Trans-activating rev protein of the human immunodeficiency virus 1 interacts directly and specifically with its target RNA. *Proc. Natl. Acad. Sci. USA* **87:**4571–4575.
4. **Daly, T., K. Cook, G. Gray, T. Maione, and J. Rusche.** 1989. Specific binding of HIV-1 recombinant rev protein to the rev-responsive element in vitro. *Nature* (London) **342:**816–819.
5. **Dingwall, C., J. Ernberg, M. Gait, S. Green, S. Heaphy, J. Karn, A. Lowe, M. Singh, M. Skinner, and R. Valerio.** 1989. Human immunodeficiency virus 1 tat protein binds trans-activation-response region (TAR) RNA in vitro. *Proc. Natl. Acad. Sci. USA* **86:**6925–6929.
6. **Ensoli, B., G. Barillari, Z. Salahuddin, R. Gallo, and F. Wong-Staal.** 1990. Tat protein of HIV-1 stimulates growth of cells derived from Kaposi's sarcoma lesions of AIDS patients. *Nature* (London) **345:**84–86.
7. **Feinberg, M., R. Jarrett, A. Aldovini, R. Gallo, and F. Wong-Staal.** 1986. HTLV-III expression and production involve complex regulation at the levels of splicing and translation of viral RNA. *Cell* **46:**807–817.

8. **Felber, B., M. Hadzopoulou-Cladaras, C. Cladaras, T. Copeland, and G. Pavlakis.** 1989. Rev protein of human immunodeficiency virus type 1 affects the stability and transport of the viral mRNA. *Proc. Natl. Acad. Sci. USA* **86:**1495–1499.

9. **Feng, S., and E. Holland.** 1988. HIV-1 tat trans-activation requires the loop sequence with TAR. *Nature* (London) **334:**165–167.

10. **Frankel, A., and C. Pabo.** 1988. Cellular uptake of the Tat protein from human immunodeficiency virus. *Cell* **55:**1189–1193.

11. **Gartenhaus, R. B., and M. E. Klotman.** 1991. Soluble Tax is taken up by CNS cells and primary lymphocytes. (Abstract) *AIDS Res. Hum. Retroviruses* **7:**229.

12. **Gaynor, R., E. Soultanakis, M. Kuwabara, J. Garcia, and D. Sigman.** 1989. Specific binding of a Hela cell nuclear protein to RNA sequences in the human immunodeficiency virus transactivation region. *Proc. Natl. Acad. Sci. USA* **86:**4858–4862.

13. **Hauber, J., A. Perkins, E. Heimer, and B. Cullen.** 1987. Transactivation of human immunodeficiency virus gene expression is mediated by nuclear events. *Proc. Natl. Acad. Sci. USA* **84:**6464–6468.

14. **Heaphy, S., C. Dingwall, I. Ernberg, M. Galt, S. Green, J. Karn, A. Lowe, M. Singh, and M. Skinner.** 1990. HIV-1 regulator of virion expression (rev) protein binds to an RNA stem-loop structure located within the rev response element region. *Cell* **60:**685–693.

15. **Kim, S., J. Byrn, J. Groopman, and D. Baltimore.** 1989. Temporal aspects of DNA and RNA synthesis during human immunodeficiency virus infection: evidence for differential gene expression. *J. Virol.* **63:**3708–3713.

16. **Laspia, M., A. Rice, and M. Matthews.** 1989. HIV-1 tat protein increases transcriptional initiation and stabilizes elongation. *Cell* **59:**283–292.

17. **Malim, M., E. Bohnlein, J. Hauber, and B. Cullen.** 1989. Functional dissection of the HIV-1 rev trans-activator—derivation of a trans-dominant repressor of rev function. *Cell* **58:**205–214.

18. **Malim, M., J. Hauber, S.-Y. Le, J. Maizel, and B. Cullen.** 1989. The HIV-1 rev trans-activator acts through a structured target sequence to activate nuclear transport of unspliced viral mRNA. *Nature* (London) **338:**254–257.

19. **Malim, M., L. Tiley, D. McCarn, J. Rusche, J. Hauber, and B. Cullen.** 1990. HIV-1 structural gene expression requires binding of the rev transactivator to its RNA target sequence. *Cell* **60:**675–683.

20. **Pavlakis, G., and C. Rosen.** Personal communication.

21. **Pavlakis, G. N., and B. K. Felber.** 1990. Regulation of expression of human immunodeficiency virus. *New Biol.* **2:**20–31.

22. **Rappaport, J., S.-J. Lee, K. Khalili, and F. Wong-Staal.** 1989. The acidic amino-terminal region of the HIV-1 tat protein constitutes an essential activating domain. *New Biol.* **1:**101–110.

23. **Rosen, C., E. Terwilliger, A. Dayton, J. Sodroski, and W. Haseltine.** 1988. Intragenic cis-acting art gene-responsive sequences of the human immunodeficiency virus. *Proc. Natl. Acad. Sci. USA* **85:**2071–2075.

24. **Roy, S., U. Delling, C.-H. Chen, C. Rosen, and N. Sonenberg.** 1990. A bulge structure in HIV-1 TAR RNA is required for tat binding and tat-mediated trans-activation. *Genes Dev.* **4:**1365–1373.

25. **Ruben, S., A. Perkins, R. Purcell, K. Joung, R. Sia, R. Burghoff, W. A. Haseltine, and C. A. Rosen.** 1989. Structural and functional characterization of human immunodeficiency virus *tat* protein. *J. Virol.* **63:**1–8.

26. **Sadaie, M. R., T. Benter, and F. Wong-Staal.** 1988. Site directed mutagenesis of two trans-regulatory genes (tat-III and trs) of HIV-1. *Science* **239:**910–913.

27. **Schwartz, S., B. K. Felber, D. M. Benko, E.-M. Fenyö, and G. N. Pavlakis.** 1990. Cloning

and functional analysis of multiply spliced mRNA species of human immunodeficiency virus type 1. *J. Virol.* **64**:2519–2529.

28. **Sharp, P., and R. Marciniak.** 1989. HIV TAR: an RNA enhancer? *Cell* **59**:229–230.

29. **Tada, H., J. Rappaport, M. Lashgari, S. Amini, F. Wong-Staal, and K. Khalili.** 1990. Transactivation of the JC virus late promoter by the tat protein of type 1 human immunodeficiency virus in glial cells. *Proc. Natl. Acad. Sci. USA* **87**:3479–3483.

30. **Vaishnav, V., M. Vaishnav, and F. Wong-Staal.** 1991. Identification and characterization of a nuclear factor that specifically binds to the REV response element (RRE) of human immunodeficiency virus type 1 (HIV-1). *New Biol.* **3**:142–150.

31. **Wong-Staal, F.** 1990. Human immunodeficiency viruses and their replication, p. 1529–1544. *In* B. N. Fields, D. M. Knipe, R. M. Chanock, M. S. Hirsch, J. L. Melnick, T. P. Monath, and B. Roizman (ed.), *Virology,* 2nd ed. Raven Press, New York.

32. **Wong-Staal, F., S. Daefler, and M. Klotman.** Unpublished data.

33. **Wong-Staal, F., and M. R. Sadaie.** 1988. Role of the two essential regulatory genes of HIV in virus replication, p. 1–10. *In* R. Franza, B. Cullen, and F. Wong-Staal (ed.), *The Control of Human Retrovirus Gene Expression.* Cold Spring Harbor Laboratory, Cold Spring Harbor, N.Y.

34. **Zapp, M., and M. Green.** 1989. Sequence-specific RNA binding by the HIV-1 rev protein. *Nature* (London) **342**:714–716.

35. **Zuker, M.** 1989. Computer prediction of RNA structure. *Methods Enzymol.* **180**:262–288.

Viruses That Affect the Immune System
Edited by Hung Y. Fan et al.
© 1991 American Society for Microbiology, Washington, DC 20005

Chapter 4

Properties of NF-κB, LBP-1, and TCF-1: Cellular Proteins That Interact with the Human Immunodeficiency Virus Type 1 Promoter in T Cells

Harald Dinter, Philip L. Sheridan, Marian L. Waterman, and Katherine A. Jones

The human immunodeficiency virus type 1 (HIV-1) promoter is a compact, highly inducible control region that can be activated by such diverse agents as the HIV-1 Tat regulatory protein, cellular signal transduction systems, and heterologous viral *trans*-activator proteins that may be found in coinfected cells (for reviews, see references 7, 21, 27, 38, 43). At least seven different proteins have been reported to bind the HIV-1 promoter, including NFAT-1 (39), NF-κB (31), EBP-1 (49), TCF-1, Sp1, LBP-1/UBP-1 (22, 49), and CTF/NF-1 (22), and in addition the functional TATA element in the LTR is contacted by the TATA factor TFIID. The arrangement of some of these control regions is outlined in Fig. 1, which serves as a reference for the three cellular proteins (NF-κB, TCF-1, and LBP-1) that are discussed in detail in this report.

THE HIV-1 PROMOTER: TCF-1 AND LBP-1

TCF-1

TCF-1 is a T-cell-specific protein, found in T lymphocytes (Jurkat) but not detected in extracts from mature B (JY, Namalwa) or nonlymphoid

Harald Dinter, Philip L. Sheridan, Marian L. Waterman, and Katherine A. Jones • The Salk Institute for Biological Studies, La Jolla, California 92037.

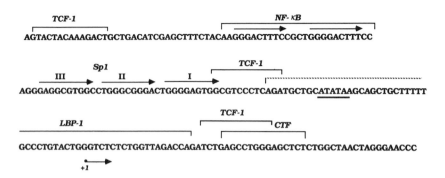

Figure 1. Arrangement of different cellular transcription factor binding sites in the HIV-1 promoter. The locations of NF-κB, TCF-1, and LBP-1 domains relative to the viral RNA start site (+1) are indicated for reference.

(HeLa) cell lines, that binds to multiple sites on the HIV-1 promoter. To purify TCF-1, extracts of uninduced Jurkat nuclei were fractionated over phosphocellulose, calf thymus DNA, and FPLC Mono Q resins. Fractions were eluted with gradients of increasing ionic strength, and TCF-1 activity was monitored by DNase I footprint experiments. The elution position of TCF-1 from these conventional columns is detailed in Fig. 2. Of these different fractionation steps, the Mono Q resin was most helpful since TCF-1 activity was found in the flowthrough fractions, whereas the bulk of the protein bound to the resin.

To purify this T-cell-specific factor further, these fractions were passed over DNA affinity columns bearing TCF-1 binding sequences from the HIV-1 promoter. Column fractions containing TCF-1 binding activity were shown by Southwestern and sodium dodecyl sulfate-polyacrylamide gel electrophoresis (SDS-PAGE) analysis to consist of a 57/53-kDa protein doublet (46). The 57- and 53-kDa TCF-1 proteins were detected in both 0.8 M and 1.5 M elution steps from the affinity resin (Fig. 3); however, the 0.8 M fraction was cross-contaminated with a series of peptides of 37 to 42 kDa. Amino acid sequence analysis of this lower group of peptides revealed the presence of a nonspecific, single-stranded DNA binding protein (45) that was unrelated to TCF-1. By contrast, the high-affinity protein fraction (1.5 M) contained only the 57/53-kDa TCF-1 proteins.

Affinity-purified TCF-1 was found to bind to three tandem sites on the HIV-1 promoter (Fig. 3), including a distal upstream site (−142 to −122) adjacent to the NF-κB site, a central site (−51 to −37) that overlaps the proximal Sp1 site, and a downstream site (+17 to +32) that overlaps the TAR element required for *trans*-activation by Tat. Ten different high-affinity TCF-1 sites mapped on the HIV-1 and other promoters (including the p56[lck] gene, CD3 γ/δ⁻ subunit genes, and TCRα enhancer) were found to share the motif 5′-PyCT(T/G)T(T/G)-3′. This core motif alone is necessary for high-affinity binding of TCF since a mutation that changes the sequence 5′-CTTTG-3′ to 5′-CATAG-3′ eliminates specific binding of TCF-1. However, the core is not sufficient for binding because it also occurs in the lymphoid-specific SL3-3 retrovirus enhancer (5, 41), which is not recognized by purified TCF-1. The role of TCF-1 in HIV promoter activity is unclear at present since mutation of the most distal site does not dramatically reduce HIV-1 gene expression in Jurkat cells (45). Thus, either TCF-1 is not a significant effector of HIV-1 promoter activity, or all three sites must be simultaneously inactivated in order to eliminate TCF-1 activity.

To investigate the binding requirements of TCF-1 in greater detail, we tested the ability of the purified protein to recognize TAR substitution mutants. Mutants +24/+27 or +30 to +33 (22) eliminated the binding of TCF-1 (Fig. 4B; compare lanes 3, 6, and 7). Transcription of these same promoter templates in a cell-free nuclear extract derived from Jurkat cells revealed that RNA initiation from the long terminal repeat (LTR) was unaffected by mutants that eliminate binding of TCF-1 (Fig. 4A). Thus, the downstream TCF-1 site can be eliminated without affecting basal viral transcription in vitro. Although these mutations also destroy Tat *trans*-activation in vivo, it is extremely unlikely that TCF-1 plays a role in this process since it does not bind avidly to the HIV-2 promoter in the TAR region, and since the TAR element is functional in RNA rather than DNA form.

TCF-1 Is Critical for TCRα Enhancer Activity in Jurkat (T) and CEM (Pre-T) Cell Lines

To better define the role of TCF-1 in transcriptional control, it was important to identify genes regulated by TCF-1. Of various T-cell-specific enhancers that have been described recently, a likely TCF-1 site was contained within the human and murine enhancers of the T-cell receptor Cα subunit gene (19, 47). This enhancer is specifically active in TCRα/β⁺ cells and consists of a tissue-specific activation region containing only three protein binding sites, flanked by extensive regions that act to silence enhancer activity in nonexpressing TCR cell lines (19, 47). One of the three activator proteins may belong to the CREB/ATF family of proteins (19); however, this particular protein differs from CREB in that it is constitutively active in Jurkat cells rather than inducible upon treatment with Forskolin or other agents that elevate cellular cyclic AMP levels (45).

A double point mutation in the TCF-1 binding site of the TCRα enhancer inactivated binding and dramatically affected enhancer activity in vivo, re-

A. Phosphocellulose gradient

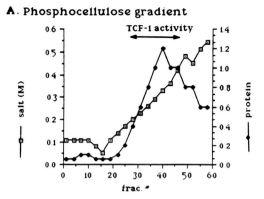

B. Calf Thymus DNA gradient

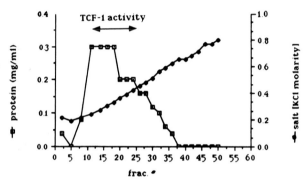

C. FPLC Mono Q gradient

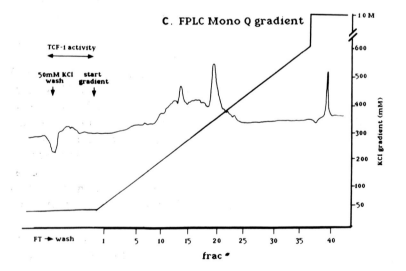

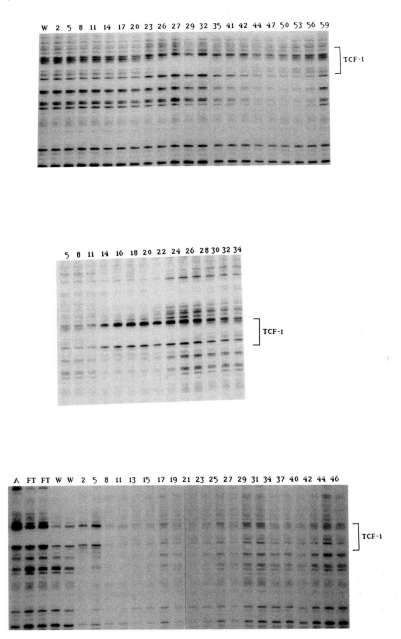

Figure 2. Chromatographic properties of TCF-1. Jurkat nuclear extracts were fractionated on phosphocellulose, calf thymus DNA-cellulose, and FPLC Mono Q columns as indicated in each panel, and fractions were pooled according to TCF-1 activity.

A . SDS-PAGE

0.8 M 1.5M   m

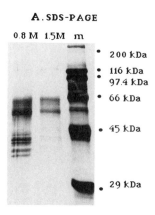

• 200 kDa

• 116 kDa
  97.4 kDa

• 66 kDa

• 45 kDa

• 29 kDa

B . DNase I footprint

142 -122   -51   -37   +17   +32

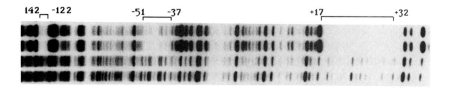

Figure 3. Purified TCF-1 binds to three discrete sites on the HIV-1 promoter. (A) SDS-PAGE analysis of affinity-purified TCF-1 (0.8 M and 1.5 M elution steps) after two sequential passes on HIV-1 DNA affinity resins. Approximately 10 μg of TCF-1 was obtained from 150 ml (150 mg) of Jurkat nuclear extract. (B) DNase I footprint analysis of the binding of purified TCF-1 to the HIV-1 promoter. Footprint boundaries of three TCF-1 sites are indicated with brackets.

vealing that TCF-1 is in fact a potent activator of gene expression (46). Thus TCF-1 functions within a 98-bp core region of the TCRα enhancer in which it is flanked by a CRE-binding protein and a second T-cell-specific protein, TCF-2. Ligated multimers of TCF-1 do not, however, have enhancer activity on their own in T-cell lines. Moreover, mutations in either the CRE or TCF-2 binding sites eliminate TCRα enhancer activity (46), indicating that TCF-1 belongs to an unusual class of transcription factors whose activity is dependent upon the context of its binding site relative to binding sites for other proteins. The observation that the T-cell-specific enhancers of the CD3-ε gene (6) and other components of the CD3/TCR complex also contain either bona fide TCF-1 binding sites or sequences related to the consensus TCF-1 binding site suggests that TCF-1 could play a wide role in the expression of genes critical to T-lymphocyte formation.

A. Primer extension   **B. DNase I Footprint**

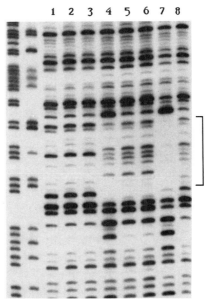

Figure 4. The downstream TCF-1 binding site does not affect HIV-1 transcription in vitro. (A) In vitro transcription reactions containing 150 ng of the wild-type HIV-1 promoter (lanes 1 and 4) or the +24/+27 (lane 2), +30/+33 (lane 3), or triple (lane 5) HIV-1 promoter mutants. Reactions were carried out as described in reference 22, and RNA was detected by primer extension. (B) Footprint analysis of the binding of TCF-1 to the wild-type HIV-1 promoter (lane 7), +24/+27 (lane 3); or +30/+33 (lane 6) mutants. Corresponding control reactions (no added protein) for each template are shown in lanes 1, 2, 4, 5, and 8.

The TCRα subunit gene is activated late during development, after the T-cell receptor β-chain genes. The immature T-cell line, CCRF-CEM, is used as a model system for this late stage of development, since the TCRβ genes are actively transcribed, whereas the TCRα enhancer is most functional following activation by phorbol ester tumor promoter (47) (e.g., TPA). In this premature T-cell line, activation of the enhancer by TPA does not result in dramatic changes in the affinity of any of these proteins for DNA, and mutational studies indicate that TCF-1 per se is not the sole target of TPA activation in CCRF-CEM cell lines.

Using amino acid sequence information obtained from the purified TCF-1 protein, we have recently cloned the gene encoding the TCF-1 transcription factor. These cDNAs will provide the reagents necessary to study the structure and function of the TCF-1 protein, understand its restriction expression in the thymus and in T-cell lines, determine its role in the expression of T-cell-specific genes, and finally establish the extent to which this factor can influence the activity of the HIV-1 promoter in T lymphocytes.

LBP-1

LBP-1 (leader-binding protein) was originally detected with experiments designed to monitor the effects of mutations within the HIV-1 promoter-proximal downstream region on viral RNA synthesis in vitro. Two mutationally sensitive regions of the leader were identified, one between −10 and +3 near the initiator and the second (the LBP-1 element) located between +3 and +20 (22). The activity of this latter region corresponded precisely with the ability of the HeLa nuclear protein, LBP-1, to bind to DNA. The effects of these mutations did not correspond with those of TAR, the element from +14 to +42 that is required for Tat *trans*-activation. In addition, LBP-1 was shown to bind to a 120-bp domain that lies downstream of the HIV-1 promoter, suggesting that the LBP-1 binding site is conserved and therefore likely to be important for some aspect of HIV gene regulation (22).

LBP-1 can be purified from nuclear extracts of the HeLa cell line by the strategy delineated in Fig. 5. LBP-1 activity is initially isolated following

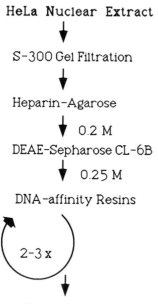

HeLa Nuclear Extract

S-300 Gel Filtration

Heparin-Agarose

  0.2 M

DEAE-Sepharose CL-6B

  0.25 M

DNA-affinity Resins

2-3 x

LBP-1 (60 kDa)

Figure 5. Strategy followed for purification of LBP-1 from HeLa nuclear extracts.

sequential S-300 and heparin-agarose columns. The protein is greatly enriched by fractionation over DEAE-Sepharose (CL-6B) resins, since LBP-1 binds at salt concentrations that prevent the attachment of many cellular DNA binding proteins. The protein is then purified to apparent homogeneity following several passes over DNA affinity columns containing ligated multimers of the LBP-1 binding site from the HIV-1 promoter. Affinity-purified LBP-1 is approximately 60 kDa in size by SDS-PAGE and silver staining (40). The DNA binding and size properties of LBP-1 are very similar to those of the UBP-1 protein described by Gaynor and colleagues (13, 14, 49), and although discrepancies exist between the reported downstream boundaries of the LBP-1/UBP-1 footprint, the two proteins are likely to be identical.

The potent effect of LBP-1 on HIV-1 transcription is readily apparent in Fig. 6, which compares the relative activity in vitro of a triple size substitution mutant promoter (triple), which eliminates the binding of purified LBP-1, with an otherwise identical wild-type HIV-1 promoter template. RNA synthesis was carried out using active extracts from either HeLa or Jurkat nuclei, and the RNA produced from each promoter was measured by the

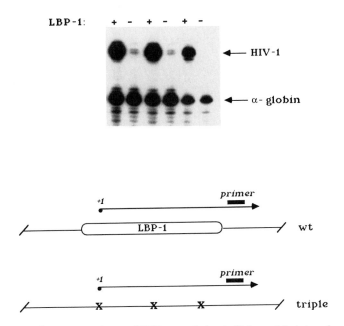

Figure 6. LBP-1 is a strong activator of HIV transcription in HeLa and Jurkat nuclear extracts. In vitro transcription reactions contained 150 ng of either the wild-type HIV-1 promoter (+) or the triple promoter mutant (−) which is incapable of binding LBP-1 (22). Reactions (30 min, 30°C) contained approximately 50 μg of either HeLa (lanes 1 through 4) or Jurkat (lanes 5 and 6) nuclear extracts, and RNA was detected by primer extension (22).

primer extension technique. In each extract, the wild-type promoter was greater than 20-fold more active than the mutant template which cannot bind LBP-1. Thus, LBP-1 is a powerful inducer of the HIV-1 LTR. Nevertheless, as indicated above, the viral Tat regulatory protein is relatively insensitive to LBP-1, indicating that Tat and LBP-1 act at different steps in the transcription reaction.

LBP-1 binding has a moderate effect on HIV RNA levels in TPA-treated cells (22) and has recently been implicated in transcriptional activation of the HIV-1 LTR by the adenovirus (26) and human cytomegalovirus (HCMV) "immediate early" proteins (3). The effects of heterologous viral *trans*-activators on HIV-1 promoter activity in vivo have been reported by several groups (8, 16, 20, 26, 29, 30, 32, 34, 36) that have examined the effects of different DNA viruses on HIV-1 gene expression. In addition, HIV and HCMV have been shown to coinfect cells in the brain (33), further suggesting that heterologous viral cofactors might have potent effects on AIDS patients. Although herpes simplex virus type 1 IE-0 protein can *trans*-activate the HIV-1 LTR in transient expression assays (29–31), Albrecht et al. (2) have recently shown that the ICP-4 (IE-3) protein is required for HIV-1 replication in T cells. Thus, more than one viral IE gene may affect the viral life cycle in coinfected cells. HCMV coinfection may be a particularly acute problem since the virus also induces expression of Fc receptors that can be used by HIV-1 to infect other cells in the body (28).

Although these heterologous viral *trans*-activators fall into distinct protein classes, several of these (e.g., the adenovirus E1A protein, pseudorabies IE, and the HCMV IE-2 [18, 35]) appear to share a common mechanism of transcriptional activation that leads to an enhancement of the efficiency of the TATA element and its cognate factor, TFIID (1, 49). The effect of the pseudorabies virus IE regulatory protein on TFIID activity has been demonstrated in vitro by Roeder and colleagues, using a system in which the templates are assembled into nucleosomes to better represent the effects of chromatin on transcription (1, 48). These findings suggest that one role for LBP-1 might be to assist in the association of TFIID with the promoter, and indeed cooperative binding of LBP-1/UBP-1 with a protein that binds over the TATA region is observed in extracts (14, 22). In concurrence with this model, we recently discovered that the LBP-1 site of the HIV-2 promoter, which is located in a more distal downstream region that does not overlap the HIV-2 TATA element and does not influence binding to TATA, is not required for *trans*-activation of the HIV-2 promoter by the HCMV IE-1,2 proteins (11). Therefore, while LBP-1 may be required for HCMV *trans*-activation of the HIV-1 LTR, its binding site is unlikely to have been conserved for the purpose of cross-activation by heterologous DNA viruses.

THE ENHANCER: NF-κB AND EBP-1

48- and 52-kDa NF-κB Proteins from JY Cells

The HIV-1 enhancer consists of a pair of 12-bp repeated elements that lie immediately upstream of the Sp1 binding sites (Fig. 1). This region of the promoter confers a substantial portion of the induction of viral RNA synthesis that is seen in TPA-treated T lymphocytes. This activation is detectable in several different cell types and involves the binding of NF-κB (31), a transcriptional activator that is expressed constitutively in mature B lymphocytes but is inducible in most other cell types by treatment with TPA (27). Although sequences similar or identical to the HIV-1 enhancer are found in the simian virus 40 enhancer core and the HCMV, β_2-microglobulin, and kappa-light chain gene enhancers, these elements can be recognized by different proteins (for review, see reference 23). For example, transcription factors AP-2 and AP-3 recognize the simian virus 40 enhancer core in uninduced HeLa cells, and DNA-binding elements for either protein are sufficient to mediate induction by TPA. A similar sequence motif in the mouse class I H-2K^b promoter is recognized by H2TF1, which may be related to the 48-kDa mouse thymoma protein KBF1. At least two other cellular proteins of distinct molecular size, HIVEN86A (12) and EBP-1 (49), have been demonstrated to bind to the HIV-1 enhancer. To unravel the role of these different proteins in transcription of HIV, it will be important to purify these individual proteins and compare their relative affinities and activities on the HIV-1 and HIV-2 promoters in vitro.

We previously showed that the HIV-1 promoter is activated in vitro by extracts derived from HeLa cells treated with phorbol esters (10). Transcriptional activation is accompanied by increased levels of a DNA binding protein, presumably NF-κB, that recognizes both HIV-1 enhancer repeats. To better understand NF-κB, we purified the protein from an Epstein-Barr virus-transformed human mature B-cell line (JY) that contains high levels of constitutively active NF-κB protein (Fig. 7) which binds to the HIV-1 enhancer in a manner indistinguishable from that seen in TPA-induced HeLa cells (10). Extracts from the JY cell line were fractionated by sequential chromatography on heparin-agarose and DNA affinity columns containing multimers of both HIV-1 enhancer repeats (-110 to -74). After three cycles over the affinity column, two proteins (52 kDa and 48 kDA) were identified by SDS-PAGE and silver staining (Fig. 7A). A final pass over the affinity column further purified these two proteins from residual contaminating proteins (Fig. 7B). This procedure yielded 0.5 to 1.0 μg of purified protein from 100 ml of JY nuclear extract (2×10^9 to 5×10^9 cells) (Table 1), and approximately 1 ng of protein generated a footprint on 2 to 5 fmol of HIV-1 enhancer DNA.

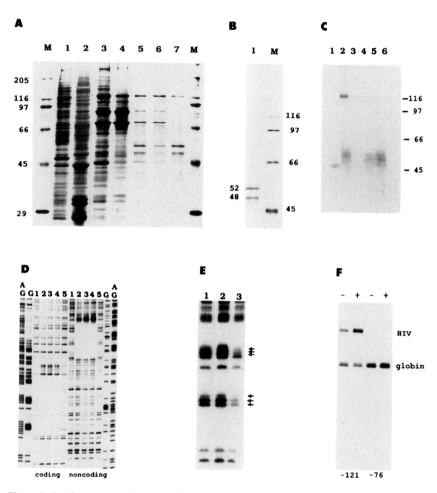

Figure 7. Purification and characterization of NF-κB from JY cells. (A) SDS-PAGE analysis of B-cell (JY) nuclear extract (0.6 footprint unit [f.u.]; lane 1), heparin agarose (50 f.u.; lane 2), and DNA-affinity chromatography: first pass (50 f.u.; lane 3); second-pass flowthrough (lane 4); second-pass eluate (50 f.u.; lane 5); third-pass flowthrough (lane 6); and third-pass eluate (50 f.u.; lane 7). (B) Fourth-pass NF-κB fractions (50 f.u.; lane 1). This procedure yielded 0.5 to 1.0 μg of purified protein from 100 ml of JY nuclear extract (2×10^9 to 5×10^9 cells). (C) Identification of NF-κB by UV-crosslink experiments. Aliquots of 0.5 ng of [^{32}P]bromodeoxyuridine-labeled HIV-1 enhancer probe were incubated with 5 f.u. of NF-κB (first-pass affinity fraction) without competitor DNA (lane 2) or with 0.5 μg of HIV-1 wild-type DNA (*dl*121; lane 3), 0.5 μg of the enhancer deletion mutant (*dl*76; lane 4), or 0.5 μg of nonspecific poly d(I-C) competitor (lane 5). A reaction containing untreated DNA is shown in lane 1, and a reaction containing only bovine serum albumin is shown in lane 6. (D) Footprint probes labeled on the coding strand (*Ava*I) or the noncoding strand (*Bgl*II) were incubated without protein (lanes 1 and 5) or in the presence of 1, 5, or 10 f.u. of third-pass affinity-purified B-cell NF-κB (lanes 2 through 4, respectively). (E)

Table 1. Purification of HIV-1 Enhancer-Binding Activity from B-Cell (JY) Extracts[a]

Step	Vol. (ml)	Protein		Activity (U)	Sp act (U/mg)	Purification factor	Yield (%)
Nuclear extract	100	350	mg	16,000	46		
0.3 M heparin agarose	28	84	mg	12,000	143	3.1	75
Affinity columns							
1	34	1.7	mg	11,000	6,470	45.2	69
2	12.5	37.5	μg	3,000	8×10^4	12	19
3	7	10.5	μg	1,250	1×10^5	1.2	7.8
4	2	0.50	μg	450	9×10^5	4.5	2.8

[a]One unit of activity is defined as the amount of protein required to protect the HIV-1 enhancer in a DNase I footprint experiment containing 2 to 5 fmol of end-labeled DNA.

A UV-crosslink experiment was carried out to confirm that the 48- and 52-kDa proteins are responsible for binding the HIV-1 enhancer (Fig. 7C). Protein from the first affinity pass was linked by UV light to a [^{32}P]bromodeoxyuridine-labeled *Ava*I-*Bgl*II fragment containing the HIV-1 promoter, digested with DNase I, and analyzed by SDS-PAGE. As shown in Fig. 7C (lane 2), autoradiography of the resulting gel revealed a doublet within the 49- to 55-kDa region of the protein gel, as well as a single band at approximately 115 kDa. Competition with *dl*121 enhancer DNA abolished binding to both the 48- and 52-kDa proteins (lane 3), whereas a deletion mutant lacking the enhancer (*dl*76) had no effect on binding of the 48- and 52-kDa proteins (lane 4). Nonspecific competitor DNA eliminated binding of the 115-kDa protein but did not affect binding of the 48- and 52-kDa proteins (lane 5), confirming that the 115-kDa protein is a nonspecific DNA-binding protein. Thus, both 48- and 52-kDa proteins bind specifically to the HIV-1 enhancer.

The binding of this purified protein spanned both enhancer repeats in footprint experiments (Fig. 7D) and generated a pattern identical to that observed previously in TPA-treated HeLa or T cells (10, 49). Binding to a deletion mutant containing only the proximal repeat resulted in a smaller

Methylation interference pattern of purified NF-κB. The *Ava*I-labeled HIV-1 DNA probe was partially methylated (lane 1) and incubated with 10 f.u. of B-cell affinity-purified protein, and bound and free fragments were isolated by gel electrophoresis for methylation interference analysis. Lane 2 shows the cleavage pattern of free DNA fragments; lane 3 shows the bound complex. (F) Purified NF-κB activates HIV-1 transcription in vitro. Transcription reactions contained 150 ng of HIV-1 DNA and 150 ng of human alpha-globin DNA and were incubated with 60 μg of HeLa nuclear extract alone (−) or extract containing 20 f.u. of the B-cell NF-κB (+), and RNA was detected by primer extension.

protected region (data not shown), indicating that the proximal repeat can be recognized independently. Methylation interference experiments were carried out to define the B-cell factor, since methylation of the most distal guanine in the kappa enhancer prevents binding of H2TF1 but does not influence binding of NF-κB (27). The results (Fig. 7E) demonstrated that methylation of the most proximal three guanines in either repeat (-104 to -102 in the A repeat, -90 to -88 in the B repeat) interfered with binding, whereas methylation of the terminal guanine residue at position -91 did not influence complex formation. The methylation pattern is therefore identical to that described for NF-κB and distinct from that reported for either AP-2 or AP-3. Moreover, addition of the purified 52/48-kDa proteins to a nuclear extract from uninduced HeLa cells resulted in a three- to fivefold activation of the HIV-1 promoter (Fig. 7F). Transcription from a control template (human alpha-globin) was not influenced, and no stimulation of the low basal activity was observed of an HIV-1 deletion template (*dl*76) that lacks the enhancer. We conclude that the B-cell 58/42-kDa proteins activate HIV-1 transcription and are probably identical to NF-κB.

The relationship between the 48- and 52-kDa proteins is unclear. They copurify in approximately equal amounts, and thus NF-κB activity could be a complex of two subunits, or the 48-kDa proteins may simply be a degradation product of the 52-kDa species. Kawakami et al. (25) recently purified a single 51-kDa NF-κB protein from Namalwa B cell extracts, using DNA affinity chromatography with the kappa enhancer sequence. This 51-kDa protein did not bind the HIV-1 enhancer in a manner similar to that described here; instead, it demonstrated a clear preference for the distal enhancer repeat (i.e., the A repeat), whereas the JY cell NF-κB fractions bind with equal affinity to both enhancer repeats, even at low protein concentrations. Therefore, the possibility remains that the 48-kDa protein is a distinct protein that combines with NF-κB to bind with high affinity to the HIV-1 enhancer repeats. Fractions derived from early passes on the HIV-1 enhancer affinity column contain a 65- to 70-kDa protein that may represent the loosely associated subunit required for inhibition of NF-κB by I-κB (4), a specific inhibitor that complexes with NF-κB in the cytoplasm of uninduced cells and delays its entry into the nucleus (for review, see reference 27).

HeLa Nuclear Extracts Contain a Distinct Enhancer-Binding Protein, EBP-1

The EBP-1 protein described by Wu et al. (49) is distinct from NF-κB and recognizes only the proximal enhancer repeat. To compare these proteins directly, we purified EBP-1 by chromatography on heparin-agarose, phosphocellulose, and enhancer DNA affinity columns (Fig. 8). Unexpectedly, we found that fractions eluting from the enhancer affinity column at low ionic strength (0.2 M KCl) generated the characteristic EBP-1 footprint, whereas the high-ionic-strength fraction (0.6 M KCl) contained a protein that bound to both enhancer repeats. The binding characteristics of this latter protein were identical to those of the JY cell NF-κB in DNase I footprint and meth-

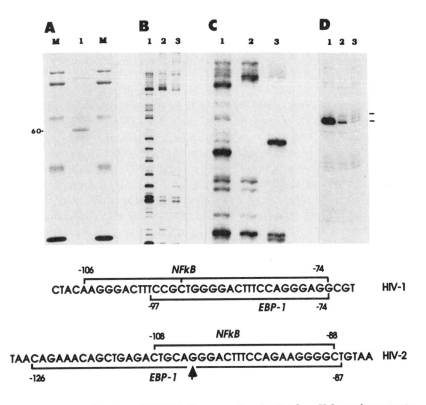

Figure 8. Affinity purification and DNA binding properties of EBP-1 from HeLa nuclear extracts. (A) 120 ng of the 0.2 M KCl EBP-1 was analyzed by SDS-PAGE. A total of 300 f.u. of NF-κB (300 ng) and 500 f.u. of EBP-1 (10 μg) were estimated to be present in HeLa extracts (2×10^{10} cells). (B) Footprint analysis of the binding of EBP-1 to the HIV-1 enhancer (*Ava*I probe), incubated without added protein (lane 1), with 1 f.u. of NF-κB (lane 2), or with 1 f.u. of the HeLa EBP-1 (lane 3). (C) Footprint analysis of the HIV-2 enhancer. The coding strand of the HIV-2 promoter was 5′-end labeled at the Asp-718 site. Reactions were carried out without added protein (lane 1), with 1.5 f.u. of NF-κB, or with 1.5 f.u. of HeLa EBP-1 (lane 3). (D) Methylation protection footprint for NF-κB (lane 1) or HeLa EBP-1 (lane 2). Two guanine residues that are differentially methylated by EBP-1 and NF-κB are indicated with arrows. The boundaries of EBP-1 and NF-κB binding sites on the HIV-1 and HIV-2 enhancers are indicated at the bottom of the figure.

ylation-protection patterns (data not shown). We conclude that HeLa nuclei contain low but detectable amounts of an activity that is similar or identical to the B-cell NF-κB and high amounts of a low-affinity protein (EBP-1) which elutes from the affinity resin at low ionic strength and binds only to the proximal enhancer repeat.

The EBP-1 fraction eluted by the 0.2 M KCl step was reapplied to the HIV-1 enhancer affinity column, and active second-pass fractions were analyzed by SDS-PAGE. The predominant protein detected in this fraction by silver staining was 60 kDa in size (Fig. 8A) and protected only the proximal (B) repeat of the HIV-1 enhancer (Fig. 8B, lane 3), whereas NF-κB consistently protected both enhancer repeats (Fig. 8B, lane 2). A difference between EBP-1 and NF-κB was also seen in methylation protection experiments (Fig. 8D). Binding of NF-κB to the coding strand of the HIV-1 enhancer protected three guanine residues (-90 to -88) in the proximal repeat, and methylation of the most distal guanine in this repeat (-91) was highly enhanced. In contrast, binding of the 60-kDa protein protected only two guanine residues (-90 and -88), and methylation of the guanine residue -91 was only slightly enhanced (Fig. 8D). Moreover, we also found that the 60-kDa EBP-1 protein and NF-κB can be differentiated by their binding to the HIV-1 enhancer (Fig. 8C), as described in greater detail below. Thus EBP-1 and NF-κB proteins have distinct DNA recognition properties as well as different apparent molecular sizes.

The most distal guanine in the B repeat of the HIV-1 enhancer is not protected against methylation by the HeLa EBP-1 protein. Furthermore, no binding of the 60-kDa EBP-1 is observed to the otherwise identical A repeat. Thus, nucleotides neighboring the B repeat appear to contribute to the binding specificity of the HeLa 60-kDa protein. Using mobility shift experiments, we determined that although the HIV-1 enhancer is recognized by both EBP-1 and NF-κB, only NF-κB formed a stable complex with the kappa enhancer (not shown), suggesting that sequences flanking the B repeat are required for efficient binding of EBP-1. Further confirmation of the difference between NF-κB and EBP-1 was obtained in in vitro transcription experiments, in which the addition of up to 20 ng of EBP-1 to either uninduced HeLa extracts or a reconstituted transcription system devoid of HIV-1 enhancer activity had no effect on HIV-1 transcription (not shown). Under these same conditions, NF-κB activated transcription up to fivefold. Similarly, no effect of EBP-1 was seen on HIV-2 promoter activity, which contains two relatively strong EBP-1 binding sites (Fig. 8C). Thus, NF-κB is a more potent transcriptional activator than EBP-1 in vitro.

Structural Arrangement of EBP-1 and NF-κB Sites on the HIV-2 Enhancer

The HIV-2 promoter is distinct from HIV-1 in that it contains only a single perfect copy of the consensus enhancer element. We were therefore interested in determining how NF-κB and EBP-1 might recognize this region. DNase I footprint experiments revealed that the high-affinity HeLa- and B-cell NF-κB proteins bound a single site on the HIV-2 promoter (−88 to −108; Fig. 8C). The sequence in this region is identical to the distal A repeat of the HIV-1 enhancer. Interestingly, the 60-kDa EBP-1 protein fraction binds two tandem sites on this region of the HIV-2 promoter. The promoter-proximal EBP-1 site maps from −87 to −105, almost completely overlapping the NF-κB site, whereas the distal EBP-1 site lies immediately upstream, from −107 to −126. Binding of EBP-1 to the two HIV-2 promoter sites occurred with greater affinity than that observed for the single site of the HIV-1 enhancer. Thus, the arrangement of EBP-1 and NF-κB sites on the HIV-2 enhancer is opposite that seen for HIV-1, consisting instead of two EBP-1 binding sites that overlap a single NF-κB site.

EBP-1 and NF-κB Recognize Distinct Elements of the HIV-2 Enhancer

To discriminate the DNA contact points of NF-κB and EBP-1, mutations in the sequence of the HIV-2 enhancer were constructed and analyzed for binding of purified fractions of each protein (Fig. 9). A single-base substitution in the NF-κB element (changing the sequence from 5'-GGGACTTTC-3' to 5'-GCGACTTTC-3'; mB) eliminates the binding of NF-κB (Fig. 9B) but does not influence the binding of EBP-1, whereas multiple mutations in the flanking region of the EBP-1 site (mC) eliminate the binding of the EBP-1 fraction but do not affect the binding of NF-κB. Analysis of heterologous promoters carrying multiple copies of each mutant enhancer revealed that whereas NF-κB DNA elements (mC) were inducible by TPA, EBP-1 DNA fragments (mB) were not (data not shown). These data confirm the observations made in vitro that EBP-1 and NF-κB are distinct transcription factors with different transcriptional properties.

NF-κB Displaces EBP-1 from Its Binding Sites on the HIV-1 and HIV-2 Enhancers

The observation that EBP-1 activity elutes from the enhancer affinity resin at low ionic strength suggested that this protein binds with lesser affinity than NF-κB and might be displaced entirely from the enhancer by NF-κB. Alternatively, these two proteins might bind simultaneously to DNA, given the closer equivalence in affinity and distinct arrangement of the NF-κB and EBP-1 sites in the HIV-2 enhancer. The interaction between affinity-purified fractions of EBP-1 (designated H, for HeLa) and NF-κB (designated B, for B-cell factor) was evaluated in DNase I footprint experiments using both the HIV-1 and HIV-2 enhancers (Fig. 10). When equivalent DNA-binding activities of EBP-1 and NF-κB were introduced at the same time to reactions

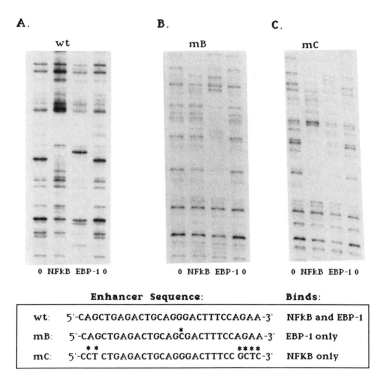

Figure 9. Mutational analysis of the NF-κB and EBP-1 binding sites on the HIV-2 enhancer. Synthetic DNA fragments of HIV-2 enhancer containing the wild-type sequence (wt) or the indicated mutations (mB, mC) were isolated and analyzed for binding-purified NF-κB and EBP-1, as indicated at the bottom of each lane.

containing fragments of the HIV-1 enhancer, only the binding pattern characteristic of NF-κB was evident (Fig. 10A). Similarly, the binding of NF-κB excluded the interaction of EBP-1 with the proximal site on the HIV-2 enhancer and strongly diminished binding to the promoter distal site (Fig. 10B), in spite of the greater affinity of EBP-1 for the HIV-2 enhancer (as compared to the HIV-1 enhancer) and the marginal overlap of the distal EBP-1 site with the single NF-κB site. We conclude that the relatively inactive EBP-1 factor cannot co-occupy the enhancer domain with NF-κB, but rather is dislodged from the HIV enhancers by the binding of NF-κB.

To test the stringency of binding site selection, the 60-kDa EBP-1 fraction was preincubated with HIV-1 DNA for 15 min prior to the addition of an equivalent amount of B-cell NF-κB, and the resultant binding pattern was determined 10 min later in a DNase I footprint experiment. For these ex-

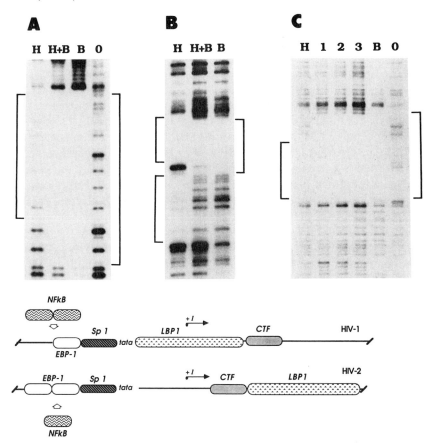

Figure 10. Purified NF-κB displaces EBP-1 from its binding site on the HIV-1 and HIV-2 enhancer in vitro. (A) HIV-1 enhancer DNA (*Ava*I probe) was incubated without protein (lane 0), with 1 f.u. of HeLa EBP-1 (H), with 1 f.u. of the B-cell NF-κB (B), or with 1 f.u. each of the EBP-1 and B-cell NF-κB (H+B) added simultaneously. (B) Binding of HeLa EBP-1 (H; 1 f.u.), the B-cell NF-κB (B; 1 f.u.), or both proteins (H+B; 1 f.u. each) to the HIV-2 enhancer (Asp-718 probe). (C) NF-κB displaces EBP-1 from the HIV enhancer regardless of whether the HeLa EBP-1 protein (H) is preincubated with the template for 15 min before B-cell NF-κB is added (B; lane 1), or both EBP-1 and NF-κB proteins are added simultaneously (lane 2), or the B-cell NF-κB is preincubated with the template prior to addition of HeLa EBP-1 (lane 3). Amounts of each protein added are as indicated for panel A.

periments, an end-labeled fragment of the noncoding strand was used to provide a clearer distinction between the two HIV-1 enhancer repeats. As shown in Fig. 10C (lane 1), the NF-κB binding pattern (lane B) predominated over the EBP-1 binding pattern (lane H) even when EBP-1 was preincubated with the DNA template. Simultaneous addition of both enhancer binding activities again resulted in complete protection over the whole enhancer region (Fig. 10C, lane 2), and as expected, EBP-1 was unable to displace NF-κB when the latter was preincubated with the HIV-1 template (Fig. 10C, lane 3). These experiments further support the conclusion that NF-κB actively displaces EBP-1 from its binding site on the enhancer and, by extension, that high levels of NF-κB in the nucleus of TPA-treated cells should dislodge the relatively inactive EBP-1 protein from the HIV-1 and HIV-2 enhancers.

Although we did not detect EBP-1 activity in low-salt affinity column elutions from B-cell extracts, the high levels of NF-κB present in these extracts may have prevented binding of EBP-1 to the resin. Indeed, EBP-1 activity can be detected in these extracts by mobility shift experiments using the mB enhancer oligonucleotide (9). The binding of EBP-1 might generate the relatively low level of HIV-1 enhancer activity observed in vivo in the absence of TPA, or this weak activity may arise from background levels of NF-κB. The identification of nonoverlapping EBP-1 and NF-κB binding site sequences will facilitate studies of the relative levels of EBP-1 activity in resting and TPA-stimulated cells. Although the purpose of EBP-1 in the uninduced cell nucleus remains unclear, the conservation of distinctly rearranged EBP-1 and NF-κB binding sites at the HIV-2 enhancer suggests a functional role for the binding of inactive EBP-1 proteins. One possibility is that binding of EBP-1 serves to prevent low levels of nuclear NF-κB from activating transcription, thereby ensuring a maximum induction response to TPA and a stringent control on basal promoter activity. Release of NF-κB from the cytoplasm to the nucleus would then displace bound EBP-1 molecules and activate the promoter. Thus the absolute response to TPA induction would vary depending upon the relative ratios of inactive EBP-1 and active NF-κB molecules in the nucleus, as well as the number and affinities of the competing EBP-1 and NF-κB binding sites at the promoter.

We conclude that a conserved biological response to T-cell activation signals is mediated by transcription control elements that are distinct in number and internal arrangement between HIV-1 and HIV-2, but nevertheless interact with at least two different cellular proteins. The scenario of events following TPA induction in T cells may be further complicated by the activation of DNA binding proteins distinct from NF-κB, including HIVEN86A, an 86-kDa protein that recognizes the HIV-1 enhancer in DNA affinity precipitation experiments using activated T-cell extracts (12), and NFAT-1 (nuclear factor of activated T cells), a protein that binds to several sites far upstream of the HIV-1 enhancer (39). It will be interesting and revealing to observe the interactions among these different enhancer-binding proteins once they are purified and analyzed directly in mixing experiments similar to the ones carried out here for EBP-1 and NF-κB.

PROSPECTUS

The data described here help to elucidate the roles of three cellular proteins with very different roles in HIV transcriptional control. TCF-1, implicated in the T-cell-specific function of the TCRα enhancer, may be an entirely context-dependent transcriptional activator whose function requires additional factors that are not found on the HIV-1 promoter. Alternatively, TCF-1 might play a role in HIV-1 expression in T cells through its central binding site, which is located in a region overlapping the proximal Sp1 binding site. LBP-1 is a particularly unusual transcription factor in that it normally activates the viral LTR from a DNA site within the untranslated leader region. It has recently been implicated in *trans*-activation of the LTR by HCMV immediate-early proteins, and indeed downstream leader elements have been previously recognized to play a role in the temporal regulation of herpesvirus genes (15). We are currently investigating the possibility that LBP-1 may recognize important control elements in herpes simplex virus type 1 and HCMV genes, and it will be important also to identify the role of this protein in the regulation of cellular genes. The observation that LBP-1 is critical for heterologous viral protein *trans*-activation but not for transcriptional induction by Tat may explain the synergism observed in vivo by the HCMV IE and Tat proteins. NF-κB, long recognized as an important activator, is shown here to displace a different, low-activity protein designated EBP-1 from both the HIV-1 and HIV-2 enhancers. Interrelationships such as these among all of the factors that bind these two promoters can best be examined with purified proteins and active in vitro transcription systems. In addition, the isolation of cDNA clones encoding these cellular factors, currently pursued in many labs, should reveal much about the different mechanisms of transcription induction.

In addition to transcriptional control by these cellular proteins, the HIV-1 and HIV-2 promoters are remarkable in their response to Tat. The mechanism of *trans*-activation by Tat is unknown and its clarification would be greatly assisted by the development of in vitro systems that recapitulate this step. Peterlin and colleagues (24, 37) have demonstrated that short viral leader transcripts are found in transfected COS cells in the absence of Tat and that Tat may function in part by overcoming a block to elongation. Transcriptional attenuation can be reproduced in vitro (42), which may help to identify the cellular proteins involved in this process. The short transcripts might accumulate the latently infected cells and damper levels of free Tat protein. These RNAs could also influence interferon-regulated pathways and could activate NF-κB directly (44). The potential of biochemical approaches to better define the function of cellular proteins in these diverse regulatory pathways may be critical for the design of effective inhibitors of viral gene expression in infected cells.

Acknowledgments. We thank Peter Barry and Paul Luciw for communicating their results on the role of the LBP-1 site in *trans*-activation by HCMV. This work was funded by grants from the National Institutes of Health, the California Universitywide Task Force on AIDS, and the Mathers Foundation. M.L.W. is supported by an American Cancer Society postdoctoral fellowship, H.D. was the recipient of a postdoctoral fellowship from the Deutsche Forschungsgemeinschaft, and K.J. is a member of the PEW Biomedical Research Scholarship Program.

REFERENCES

1. **Abmayr, S. M., J. L. Workman, and R. G. Roeder.** 1988. The pseudorabies immediate early protein stimulates in vitro transcription by facilitating TFIID: protein interactions. *Genes Dev.* **2:**542–553.
2. **Albrecht, M. A., N. A. DeLuca, R. A. Byrn, P. A. Schaffer, and S. M. Hammer.** 1989. The herpes simplex virus immediate-early protein, ICP-4, is required to potentiate replication of human immunodeficiency virus in CD4+ lymphocytes. *J. Virol.* **63:**1861-1868.
3. **Barry, P. A., and P. A. Luciw.** Personal communication.
4. **Bauerle, P. A., and D. Baltimore.** 1989. A 65-kD subunit of active NF-κB is required for inhibition of NF-κB by I-κB. *Genes Dev.* **3:**1689–1698.
5. **Boral, A. L., S. A. Okenquist, and J. Lenz.** 1989. Identification of the SL3-3 virus enhancer core as a T-lymphoma cell-specific element. *J. Virol.* **63:**76–84.
6. **Clevers, H., N. Lonberg, S. Dunlap, E. Lacy, and C. Terhorst.** 1989. An enhancer located in a CpG-island 3′ to the TCR/CD3-ϵ gene confers T lymphocyte-specificity to its promoter. *EMBO J.* **8:**2527-2535.
7. **Cullen, B. R., and W. C. Greene.** 1989. Regulatory pathways governing HIV-1 replication. *Cell* **58:**423–426.
8. **Davis, M. G., S. C. Kenney, J. Kamine, J. S. Pagano, and E. S. Huang.** 1987. Immediate-early gene region of human cytomegalovirus trans-activates the promoter of human immunodeficiency virus. *Proc. Natl. Acad. Sci. USA* **84:**8642–8646.
9. **Dinter, H.** Unpublished data.
10. **Dinter, H., R. Chiu, M. Imagawa, M. Karin, and K. A. Jones.** 1987. In vitro activation of the HIV-1 enhancer in extracts from cells treated with a phorbol ester tumor promoter. *EMBO J.* **6:**4067–4071.
11. **Duchange, N.** Unpublished data.
12. **Franza, B. R., Jr., S. F. Josephs, M. Z. Gilman, W. Ryan, and B. Clarkson.** 1987. Characterization of cellular proteins recognizing the HIV enhancer using a microscale DNA-affinity precipitation assay. *Nature* (London) **330:**391–395.
13. **Garcia, J. A., D. Harrich, E. Soultanakis, F. Wu, R. Mitsuyasu, and R. B. Gaynor.** 1989. Human immunodeficiency virus type 1 LTR TATA and TAR region sequences required for transcriptional regulation. *EMBO J.* **8:**765–778.
14 **Garcia, J. A., F. K. Wu, R. Mitsuyasu, and R. B. Gaynor.** 1987. Interactions of cellular proteins involved in the transcriptional regulation of the human immunodeficiency virus. *EMBO J.* **6:**3761-3770.
15. **Geballe, A. P., R. R. Spaete, and E. S. Mocarski.** 1986. A cis-acting element within the 5′ leader of a cytomegalovirus β transcript determines kinetic class. *Cell* **46:**865–872.

16. **Gendelman, H. E., W. Phelps, L. Feigenbaum, J. M. Ostrove, A. Adachi, P. M. Howley, G. Khoury, H. S. Ginsberg, and M. A. Martin.** 1986. Trans-activation of the human immunodeficiency virus long terminal repeat sequences by DNA viruses. *Proc. Natl. Acad. Sci. USA* **83:**9759–9763.

17. **Gentz, R., C.-H. Chen, and C. A. Rosen.** 1989. Bioassay for trans-activation using purified immunodeficiency virus tat-encoded protein: trans-activation requires mRNA synthesis. *Proc. Natl. Acad. Sci. USA* **86:**821–824.

18. **Hermiston, T. W., C. L. Malone, P. A. Witte, and M. F. Stinski.** 1987. Identification of the human cytomegalovirus immediate-early region 2 gene that stimulates gene expression from an inducible promoter. *J. Virol.* **61:**3214–3221.

19. **Ho, I.-C., L.-H. Yang, G. Morle, and J. M. Leiden.** 1989. A T-cell-specific transcriptional enhancer element 3′ of Cα in the human T-cell receptor α locus. *Proc. Natl. Acad. Sci. USA* **86:**6714–6718.

20. **Horvat, R. T., C. Wood, and N. Balachandran.** 1989. Transactivation of human immunodeficiency virus promoter by human herpesvirus 6. *J. Virol.* **63:**970–973.

21. **Jones, K. A.** 1989. HIV trans-activation and transcription control mechanisms. *New Biol.* **1:**127–135.

22. **Jones, K. A., P. A. Luciw, and N. Duchange.** 1988. Structural arrangements of transcription control domains within the 5′-untranslated leader regions of the HIV-1 and HIV-2 promoters. *Genes Dev.* **2:**1101–1114.

23. **Jones, N. C., P. W. J. Rigby, and E. B. Ziff.** 1988. Trans-acting protein factors and the regulation of eukaryotic transcription: lessons from studies on DNA tumor viruses. *Genes Dev.* **2:**267–281.

24. **Kao, S.-Y., A. F. Calman, P. A. Luciw, and B. M. Peterlin.** 1987. Anti-termination of transcription within the long terminal repeat of HIV-1 by tat gene product. *Nature* (London) **330:**489–493.

25. **Kawakami, K., C. Scheidereit, and R. G. Roeder.** 1988. Identification and purification of a human immunoglobulin enhancer-binding protein (NF-κB) that activates transcription from a human immunodeficiency virus type 1 promoter in vitro. *Proc. Natl. Acad. Sci. USA* **85:**4700–4704.

26. **Kliewer, S., J. Garcia, L. Pearson, E. Soultanakis, A. Dasgupta, and R. Gaynor.** 1989. Multiple transcriptional regulatory domains in the human immunodeficiency virus type 1 long terminal repeat are involved in basal and E1A/E1B-induced promoter activity. *J. Virol.* **63:**4616–4625.

27. **Lenardo, M. J., and D. Baltimore.** 1989. NF-κB: a pleiotropic mediator of inducible and tissue-specific gene control. *Cell* **58:**227–229.

28. **McKeating, J. A., P. D. Griffiths, and R. A. Weiss.** 1990. HIV susceptibility conferred to human fibroblasts by cytomegalovirus-induced Fc receptor. *Nature* (London) **343:**659–661.

29. **Mosca, J. D., D. P. Bednarik, N. Raj, C. A. Rosen, J. G. Sodroski, W. A. Haseltine, G. S. Hayward, and P. M. Pitha.** 1987. Activation of human immunodeficiency virus by herpesvirus infection: identification of a region within the long terminal repeat that responds to a trans-acting factor encoding by herpes simplex virus 1. *Proc. Natl. Acad. Sci. USA* **84:**7408–7412.

30. **Mosca, J. D., D. P. Bednarik, N. Raj, C. A. Rosen, J. G. Sodroski, W. A. Haseltine, and P. M. Pitha.** 1987. Herpes simplex virus type-1 can reactivate transcription of latent human immunodeficiency virus. *Nature* (London) **325:**67–70.

31. **Nabel, G., and D. Baltimore.** 1987. An inducible transcription factor activates expression of human immunodeficiency virus in T cells. *Nature* (London) **326:**711–713.

32. **Nabel, G. J., S. A. Rice, D. M. Knipe, and D. Baltimore.** 1988. Alternative mechanisms for activation of human immunodeficiency virus enhancer in T cells. *Science* **239:**1299–1302.

33. **Nelson, J., C. Reynolds-Kohler, M. Oldstone, and C. A. Wiley.** 1988. HIV and HCMV coinfect brain cells in patients with AIDS. *Virology* **165:**186–290.

34. **Ostrove, J. M., J. Leonard, K. E. Weck, A. B. Rabson, and H. E. Gendelman.** 1987. Activation of the human immunodeficiency virus by herpes simplex virus type I. *J. Virol.* **61**:3726–3732.

35. **Pizzorno, M. C., P. O'Hare, L. Sha, R. L. LaFemina, and G. S. Hayward.** 1988. Transactivation and autoregulation of gene expression by the immediate-early region 2 gene products of human cytomegalovirus. *J. Virol.* **62**:1167–1179.

36. **Rando, R. F., P. E. Pellett, P. A. Luciw, C. A. Bohan, and A. Srinivasan.** 1987. Transactivation of human immunodeficiency virus by herpesvirus. *Oncogene* **1**:13–18.

37. **Selby, M. J., E. S. Bain, P. A. Luciw, and B. M. Peterlin.** 1989. Structure, sequence, and position of the stem-loop structure in tar determine transcriptional elongation by tat through the HIV-1 long terminal repeat. *Genes Dev.* **3**:547–558.

38. **Sharp, P. A., and R. A. Marciniak.** 1989. HIV TAR: an RNA enhancer? *Cell* **59**:229–230.

39. **Shaw, J. P., P. J. Utz, D. B. Durand, J. J. Toole, E. A. Emmel, and G. R. Crabtree.** 1988. Identification of a putative regulator of early T cell activation genes. *Science* **241**:202–205.

40. **Sheridan, P. L.** Unpublished data.

41. **Thornell, A., B. Hallberg, and T. Grundstrom.** 1988. Differential protein binding in lymphocytes to a sequence in the enhance of the mouse retrovirus SL3-3. *Mol. Cell. Biol.* **8**:1625–1637.

42. **Toohey, M. G., and K. A. Jones.** 1989. In vitro formation of short RNA polymerase II transcripts that terminate within the HIV-1 and HIV-2 promoter-proximal downstream regions. *Genes Dev.* **3**:265–283.

43. **Varmus, H.** 1988. Regulation of HIV and HTLV gene expression. *Genes Dev.* **2**:1055–1062.

44. **Visnanathan, K. V., and S. Goodbourn.** 1989. Doubie-stranded RNA activates binding of NF-κB to an inducible element in the human β-interferon promoter. *EMBO J.* **8**:1129–1138.

45. **Waterman, M. L.** Unpublished data.

46. **Waterman, M. L., and K. A. Jones.** 1990. Purification of TCF-1α, a T cell-specific transcription factor that activates the T cell receptor Cα gene enhancer in a context-dependent manner. *New Biol.* **2**:621–636.

47. **Winoto, A., and D. Baltimore.** 1989. A novel, inducible and T cell-specific enhancer located at the 3' end of the T cell receptor α locus. *EMBO J.* **8**:729–733.

48. **Workman, J. L., S. M. Abmayr, W. A. Cromlish, and R. G. Roeder.** 1988. Transcriptional regulation by the immediate early protein of pseudorabies virus during in vitro nucleosome assembly. *Cell* **55**:211–219.

49. **Wu, F. K., J. A. Garcia, D. Harrich, and R. B. Gaynor.** 1988. Purification of the human immunodeficiency virus type 1 enhancer and TAR binding proteins EBP-1 and UBP-1. *EMBO J.* **7**:2117–2129.

50. **Wu, L., D. Rosser, M. C. Schmidt, and A. Berk.** 1987. A TATA box implicated in the E1A transcriptional activation of a simple adenovirus 2 promoter. *Nature* (London) **326**:512–515.

Viruses That Affect the Immune System
Edited by Hung Y. Fan et al.
© 1991 American Society for Microbiology, Washington, DC 20005

Chapter 5

Molecular Genetics of the Human Immunodeficiency Virus Type 1-CD4 Interaction

David Camerini and Irvin S. Y. Chen

CD4 is the primary high-affinity receptor for the human immunodeficiency virus type 1 (HIV-1) (20, 44), which is the etiologic agent of the acquired immunodeficiency syndrome (AIDS) (6, 32, 53). Virion binding to cell surface receptor, mediated by the *env* gene products gp120 and gp41, is the first step in the viral life cycle delimiting the cell type and species specificity of the virus. The HIV-1 external membrane glycoprotein, gp120, is highly variable from one isolate to another; this may in part be responsible for differential tropism among HIV-1 isolates (15, 26, 27, 46, 85). HIV-1/CD4 interactions may also be important for postbinding events leading to viral entry by facilitating fusion of the lipid bilayer surrounding the viral core with the plasma membrane of the target cell (62, 91). In addition, several groups of investigators have proposed that the CD4/HIV-1–*env* product interaction may be important in the pathogenesis of HIV-1 by fusing membranes within a cell or between separate cells (syncytium formation), leading to cell death (54, 56, 75, 90, 100). Recent data showing a correlation between viral pathogenesis in vivo and the ability to form syncytia in vitro lend credence to this hypothesis (15, 95, 96).

The HIV-1 *env* gene products, gp41 and gp120, form the spikes seen on the surface of HIV-1 virions in electron micrographs. HIV-1, like other re-

David Camerini and Irvin S. Y. Chen • Department of Microbiology and Immunology and Department of Medicine, University of California-Los Angeles School of Medicine, and Jonsson Comprehensive Cancer Center, Los Angeles, California 90024.

troviruses, consists of a lipid shell studded with gp120/gp41 surrounding a proteinaceous core particle composed of the *gag* gene products, which contains two copies of the viral genomic RNA and the *pol* products.

As a result of the importance of this virus-receptor pair, much effort has been devoted to the detailed study of HIV-1/CD4 interactions. Work with peptides and soluble forms of CD4 and gp120 has contributed to our knowledge of this interaction, but the most informative data have resulted from the study of mutant *env* gene products and mutant CD4 glycoproteins. In addition, for the HIV-1 *env* products, much has been learned from sequencing diverse viral isolates, since these genes are highly variable. In this review, we will summarize what is known about the structure and function of both partners in this important interaction, and we will discuss some important aspects of the HIV-1/CD4 interaction that remain unknown.

STRUCTURE AND MAPPING OF THE HIV-1 ENVELOPE GLYCOPROTEINS

The product of the HIV-1 *env* gene is the membrane protein precursor gp160, which consists of ≈ 825 amino acids after loss of the 30-residue amino-terminal signal peptide (1). This precursor is processed into the external membrane glycoprotein, gp120, of ≈ 480 residues, and the transmembrane glycoprotein, gp41, of 345 residues, by the action of a cellular protease at a conserved site following an arginine-rich sequence (79). Like some other retroviral external membrane glycoproteins, gp120 is noncovalently associated with gp41, the transmembrane glycoprotein, which is held in the lipid bilayer by virtue of a hydrophobic domain, followed by a hydrophilic region. This noncovalent association allows a significant amount of gp120 to be shed from cells (43).

gp120 and gp41 are highly glycosylated, having an average of 25 and 6 sites of N-linked lycosylation, respectively, of which 7 in gp120 and 3 in gp41 are conserved among all known HIV-1 isolates (70). In gp120, an additional seven sites are found in more than 80% of the known sequences (70). Other glycosylation sites in the hypervariable regions of gp120, while not strictly conserved, are often loosely conserved by sites within 5 to 10 residues of one another in different viral isolates. There are no O-linked glycans attached to either polypeptide (47, 64). In addition, gp120 has 18 cysteine residues conserved in all known HIV-1 isolates and gp41 has 3 cysteines, of which two are conserved (70).

The envelope gene products are the most varied proteins among HIV-1 isolates; only $\approx 65\%$ of gp120 and $\approx 80\%$ of gp41 amino acid sequence are conserved (66, 70). The sequence variation of gp120 is not uniform; some regions show only 20% sequence conservation while others show 80% conservation (66, 70). By comparing seven HIV-1 *env* gene sequences, Modrow and co-workers showed that gp120 consists of five hypervariable regions separated by less variable and conserved regions; they found no hypervariable regions in gp41 (66). Using algorithms to predict the hydrophilicity and antigenicity of gp120, they found that both parameters correlated with sequence variability among HIV-1 isolates. This, together with structural considerations, implies that the variable regions of gp120 are found on the exposed surface of the protein.

Recently, the intrachain disulfide bonds and N-linked glycans of a recombinant gp120 molecule were mapped, providing a clearer picture of the structure of gp120 (52). These investigators found that all 18 conserved cysteines of gp120 participate in intrachain sulfide bonds. These bonds are likely to be important in securing the tertiary structure of gp120, since Tschachler et al. (98) showed that substitution of other amino acids for any of seven cysteines in gp120 tested resulted in noninfectious (C-101, C-166, C-266, C-301, C-388, or C-415) or much less infectious (C-355) HIV-1 virus. The function of the oligosaccharides of gp120 is less apparent. They probably render HIV-1 less immunogenic, since they are ubiquitous on cell surface proteins. While enzymatic removal of carbohydrate in the presence of 0.025% sodium dodecyl sulfate results in 50-fold lower affinity for CD4 (60), if glycans are removed without detergent present, the affinity is reduced only 2- to 5-fold (25). Taken together, these data suggest that the carbohydrate moieties of gp120 are not directly required for the structural integrity or CD4 binding capacity of gp120, but they may stabilize its tertiary structure.

gp120

The HIV-1 external membrane protein consists of five disulfide-bound loops, which contain four of the five variable regions, and six polypeptide segments surrounding the disulfide-bound domains (Fig. 1). We shall discuss the domains of gp120 and the posttranslational modifications found by Leonard et al. on gp120 produced in Chinese hamster ovary cells, beginning at the mature amino terminus of the gp120 polypeptide and working towards the carboxy terminus (human T-cell leukemia virus type IIIb [HTLV-IIIb] isolate). The first disulfide involves cysteines at positions 24 and 44, forming a short loop. This loop is part of the relatively well-conserved amino-terminal domain of gp120, from residues 1 to 96, including a complex N-linked oligosaccharide attached to residue 58 (52). Syu et al. (94) found that deletion of residues 78 to 86 resulted in inability to bind to CD4, but they did not

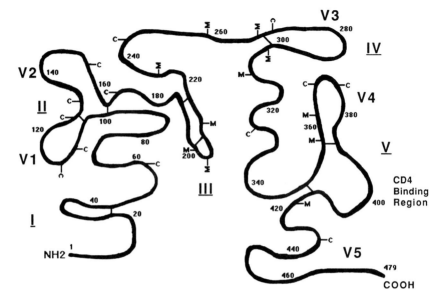

Figure 1. Representation of gp120 primary structure, simplified from reference 52, showing the disulfide bonded domains I through V, the hypervariable regions V1 through V5, and the N-linked glycans, the complex type being designated by —C and the high-mannose type by —M.

show that this was a direct effect, since they did not assay the integrity of the tertiary structure.

The second disulfide-bonded domain of gp120 is linked by two disulfide bonds between cysteines 89 and 175 and between cysteines 96 and 166 (52). The internal residues of this domain contain the first two hypervariable domains—V1, from residues 97 to 121, and V2, from residues 131 to 165—separated by a short, less variable region containing two complex N-linked glycans at residues 125 and 130 (52). V1 has two N-linked attachment sites, and V2 has one; all three are complex-type oligosaccharides (52).

The third disulfide domain is another compound loop formed by two disulfide bonds connecting cysteines 188 to 217 and cysteines 198 to 209, encompassing the sites of attachment of three high-mannose or hybrid glycans at residues 200, 204, and 211 (52). This loop forms part of the second conserved domain, which includes all amino acids between the V2 and V3 domains, and five additional conserved glycan attachment sites: complex oligosaccharides at amino acids 167 and 247 and high-mannose or hybrid glycans at residues 232, 259, and 265 (52, 66). A rabbit xenoserum directed against a peptide comprising the sequence found just C-terminal to the disulfide loop, residues 217 to 237, was capable of neutralizing diverse isolates of HIV-1 without inhibiting virion binding to CD4$^+$ cells (38). In this same region, three single amino acid changes isolated by Willey and co-workers, L-231-D, N-232-Q, and G-233-D (we use the convention of original amino acid-position-replacement amino acid and the one-letter amino acid code), resulted in loss of viral infectivity without affecting CD4 binding or syncytium formation (102). Thus, loss of the conserved N-linked glycan in the original mutant, N-232-Q, was not responsible for the phenotype observed. Interestingly, N-232-Q could be reverted by a second mutation in the V1 loop, S-98-N (102). This may indicate that the two regions are in close proximity or in some way interact in the three-dimensional structure of gp120.

The fourth disulfide domain is a singly bonded loop formed by a disulfide from C-266 to C-301, which contains the third hypervariable region (V3) defined by Modrow et al. (52, 66). The V3 loop is a major antigenic and type-specific neutralization epitope of gp120; antibodies directed to V3 block syncytium formation in an isolate- or type-specific manner, but do not affect CD4 binding (34, 59, 72, 83). The V3 loop of HTLV-IIIb has a short region of homology to the protease inhibitor trypstatin (37). During commercial preparation of gp120, cleavage at this site has been observed, but this proteolysis does not affect CD4 binding by gp120 (92). Hattori and co-workers (37) report that trypstatin can inhibit syncytium formation in infected CEM cells, suggesting that cleavage of gp120 at this site is an important step in viral entry, since syncytia are thought to form by the same process that allows virion entry.

The following interdisulfide region from amino acids 302 to 347 contains two high-mannose and one complex-type oligosaccharide at residues 302, 309, and 326, respectively (52). This region of intermediate conservation has been implicated in CD4 binding by a point mutation and a five-residue insertion described by Kowalski and co-workers (45). It is not clear whether this mutation has a direct effect on CD4 binding, or whether it may do so by disrupting the tertiary structure of gp120.

The fifth disulfide domain of gp120 is formed by two disulfides, bonding cysteines 348 to 415 and cysteines 355 to 388, and has two attached high-mannose glycans followed by two of the complex type at residues 356, 362, 367, and 376, respectively (52). The domain consists of a variable region, residues 348 to 362, followed by the fourth hypervariable region, residues 363 to 384, and a constant region of amino acids 385 to 415 (66). This constant region includes the most convincingly demonstrated CD4-binding domain, from residues 389 to 407 (18, 45, 50). The first implication of this region in CD4 binding was provided by Lasky and co-workers, who isolated a 44-amino-acid hydrolysis fragment of gp120, residues 383 to 426, by its affinity for a monoclonal antibody that blocked CD4 binding (50). They showed that this peptide could block CD4 binding to gp120 when present at 1,000 times the dissociation constant for native gp120. Subsequently, this group isolated a deletion of amino acids 396 to 407 that completely blocked CD4 binding, and a point mutation, A-403-D, that had a partial effect. Kowalski and co-workers inserted the sequence GINSG into gp120 after residue 389, which was changed from R to T (45). This perturbation also resulted in loss of CD4 binding. Cordonnier and colleagues made two short deletions in this region, residues 389 to 397 and 390 to 396, which resulted in loss of CD4 binding (18). They made a series of amino acid replacements for W-397; substitution of S, G, V, or R led to complete loss of CD4 binding, while replacement with the large hydrophobic residues Y or F had only a partial effect on CD4 binding. Mutations at position 390 also had interesting effects: I-390-K partially blocked CD4 binding, while I-390-T abrogated infection of U937 but not normal T-cells. Monoclonal antibodies directed to this region also blocked CD4 binding of and syncytium formation induced by several viral isolates (93). While the sum of all this work strongly implicates gp120 residues 389 to 407 in CD4 binding, it is still possible that this region is indirectly involved in CD4 binding by affecting the conformation of another region of gp120 that makes direct contact with CD4.

The C-terminal 64 residues of gp120, 416 to 479, are devoid of cysteines, but include a conserved high-mannose glycan at N-418 and a hypervariable region from residues 430 to 440 that contains a complex-type oligosaccharide. Kowalski et al. inserted 5 or 10 amino acids after residue 429 and found that this abrogated CD4 binding. In addition, a deletion of gp120 sequence after residue 428 with a six-residue replacement showed no CD4 binding (45). However, these mutants were not assayed for retention of tertiary structure.

The external membrane protein of HIV-1 is a polypeptide of $\approx$480 amino acids composed of five disulfide-bonded domains, four highly variable regions, and an average of 25 N-linked glycans. Thus, the exposed surface of gp120 is likely composed mostly of carbohydrate and an amino acid sequence that is not constrained by structural considerations. The result is that HIV-1, which is coated with gp120, is not very immunogenic and is able to evade recognition by mutations in the *env* gene that do not affect viral function. Since broadly neutralizing anti-HIV-1 sera are rare, the constant regions critical for gp120 function, such as the CD4 binding site, may be shielded from immune recognition by carbohydrate or variable peptide moieties.

Mutations that result in loss of CD4 binding have been isolated in four separate regions of gp120. While it is possible that all four regions make direct contact with CD4, it is more likely that some of these mutations affect the tertiary or quaternary structure of gp120 and thus render it unable to bind CD4. Resolution of this quandary awaits cocrystallization of the gp120/CD4 complex or availability of a number of anti-gp120 monoclonal antibodies (MAbs) to show that these mutations do not affect gp120 structure at distant sites. Nevertheless, the preponderance of data suggest that the region first defined by Lasky et al. (50) and later by Kowalski et al. (45) and Cordonnier et al. (18), from amino acids 389 to 407, is the site of direct contacts with CD4. The data of Cordonnier and co-workers are particularly compelling; they created several different amino acid replacements for W-397 that had graded effects on CD4 binding.

Two groups have shown that several of the C-terminal positively charged amino acids of gp120 are important for cleavage of the precursor gp160 into gp120 and gp41 (9, 35). Mutation of the normal C-terminal residue of gp120, R-481 to S or T, resulted in no cleavage, while nearby mutations R-478-5 or K-480-N resulted in less than 5% cleavage of gp160. These groups also showed that mutations which blocked normal cleavage of gp160 produced noninfectious virus (9, 35). Nevertheless, gp160 was glycosylated, reached the cell surface, and could bind CD4 (35).

The external glycoprotein of HIV-1, gp120, is a complex, highly variable molecule that is half composed of carbohydrate by mass. This protein, composed of five disulfide-bonded domains, imparts the CD4 receptor specificity to HIV-1 since it binds to CD4 with a dissociation constant in the nanomolar range for the strains tested (4, 11, 50, 51). The CD4 binding site on gp120 has been incompletely defined so far, although one region has been implicated. The variation seen in gp120 among HIV-1 isolates is likely to be a major determinant in the strain-to-strain variability in cell tropism and pathogenicity.

gp41

Much less is known about the structure of gp41 than that of gp120. This 345-amino-acid transmembrane protein has two hydrophobic regions, one at its amino terminus and one at residues 178 to 199, but the number of times it traverses the membrane is not known, nor is the disposition of each portion of the protein, be it intracellular or extracellular (68, 77, 86, 101). The amino-terminal hydrophobic region of gp41 has been implicated in syncytium formation. Kowalski and co-workers observed that insertion of four or six amino acids after residue 7 resulted in loss of HIV-1-mediated syncytium formation without affecting CD4 binding (45). Recently, Freed et al. (29) made a series of mutations in the amino terminus of gp41. They found that replacement of hydrophobic residues with charged residues such as A-1-E, V-2-E, L-9-R, A-15-E, L-26-R, or V-28-E resulted in loss of syncytium induction in HeLa-T4 cells. The charged residues following this hydrophobic sequence are also important, as a Q-29-L replacement also resulted in a non-syncytium-inducing gp41, and R-31-G had a partial effect (29). Bosch et al. (8) corroborated these results with a series of mutations in the amino terminus of the equivalent transmembrane protein of the simian immunodeficiency virus (SIV). They found that the hydrophobicity of this domain correlated directly with syncytium-inducing capacity of SIV *env* products expressed in CD4+ cell lines (8). These findings are not without precedent, in that fusion proteins of paramyxoviruses have hydrophobicity profiles and amino-terminal sequences similar to those of HIV-1 gp41 and these proteins are known to mediate membrane fusion (31).

Recently there have been several reports that gp41 and the gp41-gp120 complex exist as dimers and tetramers (22, 74). These reports suggest that the physiologic interaction of HIV-1 and CD4 may involve multimeric protein complexes. CD4 may also exist as a dimer (see below), but the stoichiometry of gp120-CD4 binding in vivo is not known.

Thus, gp41 is likely required for viral entry and syncytium formation and indirectly for CD4 binding, since it secures gp120 on the virion or infected-cell surface. However, the mechanism by which this amphiphilic protein mediates cell-cell and virion-cell fusion is unknown at present.

CD4 INTRODUCTION

CD4 is a 55×10^3-molecular-weight glycoprotein found on the surface of about two-thirds of peripheral blood T-cells (78) and at a lower level on myeloid cells (67, 103). CD4-positive T-cells recognize antigen in the context of class II MHC molecules and are central to many immune defense mechanisms (7, 23, 48, 63). The CD4-bearing T-cell subset contains the majority of T-helper cells. These cells are required for mounting an effective antibody response to extracellular pathogens, for maturation of cytotoxic T-cells that combat intracellular pathogens, and for activating phagocytic cells which engulf and destroy invading microbes. HIV-1 destroys these cells in vitro, and this may be the primary cause of AIDS in HIV-1-infected individuals. In fact, the number of CD4+ T-cells may be the best clinical indicator of disease progression in HIV-1-infected individuals (24).

CD4 is part of the T-cell antigen receptor complex. Cocapping and comodulation data suggest that CD4 associates with the T-cell antigen receptor (TCR) and is an active part of the signal transduction mechanism (3, 80, 81). Similar conclusions may be drawn from the fact that anti-CD4 MAbs block T-cell proliferation driven by some anti-TCR MAbs (36, 42). Recently, CD4 has been shown to bind major histocompatibility complex (MHC) class II antigens and to be associated with the lymphocyte-specific tyrosine kinase p56^{lck} (82, 99). Cross-linking CD4 leads to activation of the kinase activity of p56^{lck} and phosphorylation of CD3ζ, part of the TCR complex. Thus, a coherent view of the TCR includes $\alpha\beta$ or $\gamma\delta$ variable chains associated with the CD3 complex γ, δ, ϵ, and ζ and with either CD4 or CD8 plus p56[lck]. These latter molecules determine MHC class specificity and contribute to signal transduction, respectively (41, 69).

CD4 STRUCTURE AND MAPPING

CD4 is a type 1 membrane glycoprotein of the immunoglobulin gene family (57). Nascent CD4 has a hydrophobic 24-amino-acid leader peptide not found in mature CD4, whose amino terminus is known from sequencing of soluble forms (27, 39). The mature CD4 glycoprotein is predicted to have 375 extracellular amino acids which comprise four immunoglobulin homologous domains, a 21-residue membrane-spanning domain, and a 39-residue cytoplasmic tail which is highly conserved (Fig. 2).

The amino-terminal domain 1, residues 1 through 109, is homologous to immunoglobulin variable regions and is the region which interacts with HIV-1 (17, 49, 65, 73). Most of the anti-CD4 MAbs that have been mapped react with domain 1; all but one react with domain 1 or domain 2 or both together (73, 87). Epitope mapping and sequence alignment of domain 1 of CD4 suggest that it has the folding pattern of an immunoglobulin variable region. Many workers have used an alignment of CD4 domain 1 with the human immunoglobulin kappa REI variable region to show that CD4 regions homologous to surface residues of the REI crystal structure are likely to be on the surface of CD4, as they affect binding of gp120 or anti-CD4 MAbs (4, 10, 11, 73, 87). MAbs, whose binding is blocked by several separate mutations at sites that are distant in the primary sequence but are predicted to be adjacent in an immunoglobulin fold, provide further evidence for the structural similarity of CD4 domain 1 with immunoglobulin variable regions (87).

Molecular genetic studies indicate that part of the amino-terminal immunoglobulinlike domain of CD4, corresponding to CDR2, the second complementarity-determining region of the antibody variable region, interacts

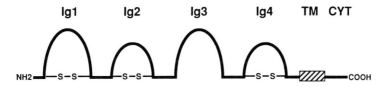

Figure 2. Representation of CD4 structure showing the immunoglobulin (Ig) homologous domain Ig1 through Ig4, the transmembrane domain (TM), and the cytoplasmic domain (CYT).

with HIV-1 (4, 17, 49, 65, 73, 87). Many separate single amino acid replacements severely restricting HIV-1 binding have been identified in the CDR2 homologous region: G-41-S, S-42-Y or -L, F-43-V, L-44-S, T-45-P or -G, G-47-R or -S, S-49-Y (4, 10, 73). The binding site for HIV-1 in its entirety may span residues 41 through 67 (4, 10). Most of the mutations do not affect the recognition of CD4 by a large panel of anti-CD4 MAbs directed to other epitopes in the first immunoglobulin V-like domain, indicating that this region is likely to be in direct contact with gp120. Recent work implicates the region of CD4 homologous to the third complementarity-determining region of an immunoglobulin molecule in HIV-1-mediated syncytia formation (11). Mutations affecting syncytium formation (E-87-G, D-88-N, and Q-89-K or -L) were found near the beginning of the CD4 segment analogous to CDR3, predicted to be spatially proximate to the major binding site in CDR2. Again, these mutations do not affect the binding of other anti-CD4 MAbs, indicating that the overall structure is likely preserved (11). Synthetic CD4 peptides corresponding to a portion of CDR2 (40) or CDR3 (55, 88) have also been reported to inhibit syncytium formation and block viral infectivity; however, the concentrations at which the peptides show activity are four to five orders of magnitude higher than the dissociation constant for HIV-1 binding to native CD4.

All of the extracellular variation between chimpanzee and human CD4 resides in domain 1 (11). These changes include two unique sites of N-linked glycosylation and a divergence in the CDR3-homologous domain that was found to be important in HIV-1-mediated syncytium formation (11). (Nevertheless, these domains are 96% homologous.) Perhaps this divergence reflects selection for maintained recognition of MHC molecules, which are known to have higher-than-average mutation rates (61). In contrast, the rhesus monkey, rat, and mouse CD4s display roughly average extracellular homology with the human sequence in this domain: 89%, 50%, and 53%, respectively (11, 16, 58, 97).

The second immunoglobulinlike domain, residues 110 through 179, is homologous to immunoglobulin constant regions, although it is truncated, leaving 28 residues between the two cysteines rather the characteristic 60 (16). Domain 2 was originally thought to play a role in HIV-1 binding on the basis of reduced gp120 binding by human/mouse CD4 chimeras (17, 49). However, more recent work suggests that this is not the case, since soluble CD4 domain 1 proteins bind HIV-1 with nearly the same affinity and block HIV-1-induced syncytia nearly as well as full-length soluble CD4 (4, 14). Rhesus, rat, and mouse CD4s are most divergent from their human homolog in domain 2. The rhesus monkey CD4 sequence is 86% homologous, the rat sequence is 46% homologous, and the murine sequence is 50% homologous to human CD4 in this region (11, 16, 58, 97).

Domain 3 of CD4 is an immunoglobulin variable region homologous domain, although it lacks cysteines altogether (16). This is the most conserved extracellular domain between human and rhesus (98%), rat (58%), and mouse (57%) CD4s (11, 16, 58, 97), suggesting that evolutionarily constraining interprotein contacts, perhaps to a constant portion of the TCR:CD3 complex or between CD4s in a putative dimer (84), may be made by this domain. Domain 3 also contains one of the two N-linked glycan sites found in all known CD4 sequences.

Membrane-proximal domain 4 is related to immunoglobulin constant regions, though shortened to 41 residues between cysteines. In the rhesus monkey, rat, and mouse, this domain, like domain 1, shows average sequence homology compared to the other external domains (89%, 52%, and 52%, respectively) (11, 16, 58, 97). Domain 4 contains the second site for the addition of N-linked glycans that is shared by all known CD4 polypeptides, plus an additional site present only in the mouse and rat. Sattentau et al. (87) mapped the binding of the prototype CD4 MAb, OKT4, to this domain. Previous work has shown that a polymorphism in human CD4 exists with regard to the OKT4 binding site; this polymorphism was found in 30% of Black Americans in one study (30) and has no apparent effect on immune function or on HIV-1 infection (28).

The membrane-spanning domain is well conserved between human and rhesus CD4s (95%), although rodent CD4s show no bias towards conservation in this region; the rat domain is 50% homologous and the mouse transmembrane sequence is 52% homologous to the human membrane-spanning domain (11, 16, 58, 97).

The cytoplasmic domain is the most highly conserved region among the known CD4 glycoproteins. Recently, this domain of CD4 was shown to bind the tyrosine kinase p56^{lck} (89), thus explaining the evolutionary conservation observed. The rhesus macaque CD4 sequence is the same as the human in this region; the rat and mouse CD4 sequences are 79% and 80% homologous, respectively (11, 16, 58, 97). Chimpanzee CD4 is anomalous in this regard, since its only divergence from human CD4, apart from four in domain 1, is in the cytoplasmic domain at position 405 (11). Nevertheless, the cytoplasmic domain of the chimp is 97% homologous to its human counterpart.

The CD4 glycoprotein, an essential element of the human immune system, is a high-affinity receptor for HIV-1. The immunodominant amino-terminal domain of CD4, which has significant structural homology to immunoglobulin variable domains, is the HIV-1 interaction region. Regions within this domain that are homologous to two of the three antigen-binding variable loops of an immunoglobulin molecule, CDR2 and CDR3, interact with HIV-1.

AREAS FOR FUTURE STUDY

Much remains to be learned about HIV-1/CD4 interactions in terms of
viral tropism, postbinding events leading to viral entry, and differential sensitivity of HIV-1 isolates for neutralization by soluble CD4. Viral tropism is
more complicated than it initially seemed. Although CD4 is the primary
receptor for HIV-1, and certain CD4 MAbs such as anti-Leu-3a block HIV-
1 infection of virtually all susceptible cells, several lines of evidence suggest
that CD4 is not the only factor restricting the tropism of HIV-1. First, species
tropism cannot be explained by CD4 alone since human CD4-bearing rodent
and African green monkey cells bind HIV-1 but are refractory to infection
(12, 57). These same cells produce HIV-1 when transfected with infectious
clones, suggesting that the block is at the level of viral entry.

The species specificity of HIV-1 may be determined by a ubiquitous
human- and higher primate-specific protein or epitope required to allow the
cell surface fusion event. Or, of course, humans may lack a protein that blocks
this process. Such a molecule may interact with CD4, or it may interact
directly with HIV-1 gp120 or gp41. The exact function of such a second
receptor is speculative at this point, but several possibilities should be mentioned in light of the available data. These may be viewed in the context of
a three-stage model of HIV-1 infection as presented below. Perhaps cleavage
at the trypstatin homologous site in the V3 loop by a specific human enzyme
is required to activate the fusion capacity of gp41, which then allows viral
entry. It has also been suggested that gp120 may be released from gp41 upon
CD4 binding. This would then open new domains of gp41 for contact with
other cell surface proteins required for internalization. Qureshi et al. reported
such a protein that binds a peptide fragment of gp41 (76).

An observation related to the species specificity of HIV-1 is that syncytia
occur between human and rodent or African green monkey cells, where one
cell type bears human CD4 and the other expresses HIV-1 gp120 and gp41
only when the CD4-bearing cell is of human origin (5). Thus, HIV-1-mediated
syncytium formation in model systems is highly correlated with susceptibility
to HIV-1 infection. Nevertheless, this correlation is not perfect since chimpanzee T-cells are infectable but form syncytia much less readily than human
T-cells (11), and some strains of HIV-1 do not readily induce syncytia. One
explanation for this imperfect correlation comes from the finding of Layne
et al. (51) that multiple gp120/CD4 contacts are required for viral infection.
It may be that such multiple contacts are more easily formed between a virion
and a CD4-bearing cell than between an infected cell and a receptor-bearing
uninfected cell, due to a greater density of gp120 on the virion surface.

The second anomaly regarding CD4 and HIV-1 tropism pertains to
strains of HIV-1 that are apparently limited at the level of viral entry in

their ability to infect human CD4-bearing cells. Two molecularly cloned HIV-1 isolates from a single patient exemplify this limited tropism (46). HIV-1$_{JR-CSF}$, like many primary HIV-1 isolates, can only efficiently infect stimulated primary T-cells. It does not infect leukemic T-cell lines, although many bear higher levels of CD4 than primary T-cells, and it infects mononuclear phagocytes inefficiently. HIV-1$_{JR-FL}$ efficiently infects primary T-cells and macrophages, but again will not infect CD4$^+$ T-cell lines. Recombinants made between these viruses and HIV-1$_{NL4-3}$, which infects cell lines but not macrophages, has shown that both of these specificities are likely to be conferred by the envelope gene and to act upon viral entry (13, 71). One simple hypothesis to explain this tropic difference invokes a second cellular protein required for viral entry after the high-affinity CD4-gp120 binding event occurs. Such a second receptor or entry facilitator might function as hypothesized above, but must exist in several forms, or there may be several alternative second receptors such that different envelope gene products would interact productively with different second-receptor or facilitator molecules. This hypothesis is consistent with CD4-dependent infection if HIV-1 binding to CD4 is the primary event that initiates infection (see our three-stage model of HIV-1 infection below).

The third surprise with regard to HIV-1 and CD4 is the inefficiency of soluble CD4 (sCD4) in neutralizing clinical isolates of HIV-1 in vivo and in vitro. Several groups have found that clinical strains are 100- to 1,000-fold less sensitive to neutralization by sCD4 than are prototypical laboratory strains with broad host range, such as HTLV-IIIb (19, 33). This insensitivity to sCD4 seems to correlate with the limited tropism discussed above: HIV-1$_{JR-CSF}$ and HIV-1$_{JR-FL}$ are as insensitive to sCD4 as uncloned clinical isolates, most of which only infect primary T-cells. Furthermore, adaptation of a primary HIV-1 isolate to growth in a T-cell line was accompanied by an increased sensitivity to sCD4 neutralization (19). Perhaps more surprising is the report by Allan et al. (2) that sCD4 can enhance the infectivity of SIV.

We propose a three-stage model of infection by HIV-1 and related viruses, which can account for the foregoing data. First, HIV-1 binds to cell surface CD4 with an affinity that may vary by several orders of magnitude from strain to strain. Eventually, multiple contacts form between gp120 on the virion surface and CD4 on the cell surface, bringing the viral lipid envelope in close apposition to the cytoplasmic membrane. Second, an unstable membrane fusion intermediate forms, due to conformational changes in gp41 that may also involve gp120 or CD4 or both. Fusion intermediate formation may occur after or accompanied by cleavage and/or shedding of gp120, a process that may require cellular or serum factors. Third, the virion membrane fuses with the cytoplasmic membrane, injecting the viral core particle into the cytoplasm in a process that requires one or more cellular factors exclusive of CD4.

This model can account for the species specificity of all HIV-1 isolates and the cell type specificity of some isolates by the requirement for one or more second receptors or entry facilitator molecules. There must be a requirement for one such factor by all HIV-1 strains, since no known strain will infect murine or African green monkey cells bearing human CD4. The inability of many HIV-1 strains to infect certain $CD4^+$ human cells such as mononuclear phagocytes or leukemic T-cell lines may reflect a requirement for a subset of the possible facilitator molecules or an additional factor requirement. We propose that these second receptor molecules are likely to act at the second or third step of viral entry in our model by stabilizing the membrane fusion intermediate or by actively participating in the membrane fusion process.

The variability in sensitivity to neutralization by sCD4 among HIV-1 strains and the related virus, SIV-1, may result in part from differing affinities of viral gp120 molecules for CD4, but this may also result from variation in stability of the membrane fusion intermediate. According to this view, SIV can be activated by sCD4, although it likely has lower affinity for human CD4 than do HIV-1 strains because it may have a relatively stable fusion intermediate such that interaction with sCD4 prepares the virion for entry into any cell it encounters within a finite period of time. HTLV-IIIb, however, is highly sensitive to neutralization by sCD4, according to this model, because the membrane fusion intermediate that forms after interaction with sCD4 is too unstable to allow time for the virion to encounter a cell before the intermediate decays irreversibly. By this view, strains of HIV-1 that are less sensitive to neutralization by sCD4, such as those described by Daar et al. (19) and Gomatos et al. (33), may simply have lower affinity for CD4 than does HTLV-IIIb or other prototypical HIV-1 strains and thereby require higher concentrations of sCD4 for neutralization. It is also possible that they have a more stable membrane fusion intermediate than laboratory strains, but less stable than that of SIV. If this is the case, high concentrations of sCD4 would be required for neutralization to ensure that sCD4 interaction takes place quickly after viral budding to preclude the active intermediate interacting with a susceptible cell.

SUMMARY

The interaction of HIV-1 with its high-affinity cell surface receptor, CD4, is a primary event in viral infection. Sites of interaction have been mapped on both gp120 and CD4, but much remains to be elucidated. We will gain a more detailed understanding of CD4/gp120 contact when crystal structures of these molecules, and ideally of the CD4/gp120 complex, are solved. Nevertheless, the localization of CD4 mutants that affect gp120 interaction in the 109-amino-acid first immunoglobulin homologous domain of CD4 allows a reasonable prediction of the structure of CD4 domains that interact with gp120. The same cannot be said for gp120; while we now know the sites of the disulfide bridges and N-linked oligosaccharides, the size and complexity of this molecule eschew structural predictions. Subsequent to receptor binding, the cell surface fusion of virion and cell may require other cellular proteins or may result solely from conformational changes of gp120/gp41 and CD4. The amino terminus of gp41 is important for membrane fusion, and cleavage of the precursor gp160 is probably required, but the exact mechanism remains obscure. A similar cell-cell fusion event, mediated by gp120/gp41 on the surface of infected cells and CD4 on uninfected cells, is responsible for syncytium formation. This process is cytopathic and has been correlated with viral pathogenesis in vivo.

REFERENCES

1. **Allan, J. S., J. E. Coligan, F. Barin, M. F. McLane, J. G. Sodroski, C. A. Rosen, W. A. Haseltine, T. H. Lee, and M. Essex.** 1985. Major glycoprotein antigens that induce antibodies in AIDS patients are encoded by HTLV-III. *Science* **228:**1091–1094.

2. **Allan, J. S., J. Strauss, and D. W. Buck.** 1990. Enhancement of CD4 infection with soluble receptor molecules. *Science* **247:**1084–1088.

3. **Anderson, P., M. L. Blue, and S. F. Schlossman.** 1988. Evidence for a specific association between CD4 and approximately 5% of the CD3:TCR complexes on helper T lymphocytes. *J. Immunol.* **140:**1732–1737.

4. **Arthos, J., K. C. Deen, M. A. Chaikin, J. A. Fornwald, G. Sathe, Q. J. Sattentau, P. R. Clapham, R. A. Weiss, J. S. McDougal, C. Pietropaolo, R. Axel, A. Truneh, P. J. Maddon, and R. W. Sweet.** 1989. Identification of the residues in human CD4 critical for binding of HIV. *Cell* **57:**469–481.

5. **Ashorn, P. A., E. A. Berger, and B. Moss.** 1990. HIV envelope glycoprotein/CD4 mediated fusion of non-primate cells with human cells. *J. Virol.* **64:**2149–2156.

6. **Barré-Sinoussi, F., L. Montagnier, C. Dauguet, J. C. Cherman, F. Rey, M. T. Nugeyre, C. Axler-Blin, J. Gruest, S. Chamaret, F. Vézinet-Brun, C. Rouzioux, and W. Rozenbaum.** 1983. Isolation of a T-lymphotropic retrovirus from a patient at risk for acquired immune deficiency syndrome (AIDS). *Science* **220:**868–870.

7. **Biddison, W. E., P. E. Rao, M. A. Talle, G. Goldstein, and S. Shaw.** 1982. Possible involvement of OKT4 molecule in T cell recognition of class II HLA antigens. *J. Exp. Med.* **156:**1065–1076.

8. **Bosch, M. L., P. L. Earl, K. Fargnoli, S. Picciafuoco, F. Giombini, F. Wong-Staal, and G. Franchini.** 1989. Identification of the fusion peptide of primate immunodeficiency viruses. *Science* **244:**694–697.

9. **Bosch, V., and M. Pawlita.** 1990. Mutational analysis of the human immunodeficiency virus type 1 *env* gene product proteolytic cleavage site. *J. Virol.* **64:**2337–2344.

10. **Brodsky, M. H., M. Warton, R. M. Myers, and D. R. Littman.** 1990. Analysis of the site in CD4 that binds to the HIV envelope glycoprotein. *J. Immunol.* **144:**3078–3086.

11. **Camerini, D., and B. Seed.** 1990. A CD4 domain important for HIV-mediated syncytium formation lies outside the principal virus binding site. *Cell* **60:**747–754.

12. **Camerini, D., and B. Seed.** Unpublished data.

13. **Cann, A. J., J. A. Zack, A. S. Go, S. J. Arrigo, Y. Koyanagi, P. L. Green, Y. Koyanagi, S. Pang, and I. S. Y. Chen.** 1990. Human immunodeficiency virus type 1 T-cell tropism is determined by events prior to provirus formation. *J. Virol.* **64:**4735–4742.

14. **Chao, B. H., D. S. Costopoulos, T. Curiel, J. M. Bertonis, P. Chisholm, C. Williams, R. T. Schooley, J. J. Rosa, R.A. Fisher, and J.M. Marganore.** 1989. A 113-amino acid fragment of CD4 produced in *Escherichia coli* blocks human immunodeficiency virus-induced cell fusion. *J. Biol. Chem.* **264:**5812–5817.

15. **Cheng-Mayer, C., D. Seto, M. Tateno, and J. A. Levy.** 1988. Biologic features of HIV-1 that correlate with virulence in the host. *Science* **240:**80–82.

16. **Clark, S. J., W. A. Jefferies, A. N. Barclay, J. Gagnon, and A. F. Williams.** 1987. Peptide and nucleotide sequences of rat CD4 (W3/25) antigen: evidence for derivation from a structure with four immunoglobulin-related domains. *Proc. Natl. Acad. Sci. USA* **84:**1649–1653.

17. **Clayton, L. K., R. E. Hussey, R. Steinbrich, H. Ramachandran, Y. Hussain, and E. Reinherz.** 1988. Substitution of murine for human CD4 residues identifies amino acids critical for HIV-gp120 binding. *Nature* (London) **335:**363–366.

18. **Cordonnier, A., L. Montagnier, and M. Emerman.** 1989. Single amino-acid changes in HIV envelope affect viral tropism and receptor binding. *Nature* (London) **340:**571–574.

19. **Daar, E. S., X. L. Li, T. Moudgil, and D. Ho.** 1990. High concentrations of recombinant soluble CD4 are required to neutralize primary HIV-1 isolates. *Proc. Natl. Acad. Sci. USA* **87:**6574–6578.

20. **Dalgleish, A. G., P. C. L. Beverley, P. R. Clapham, D. H. Crawford, M. F. Greaves, and R. A. Weiss.** 1984. The CD4 (T4) antigen is an essential component of the receptor for the AIDS retrovirus. *Nature* (London) **312:**763–767.

21. **Doyle, C., and J. L. Strominger.** 1987. Interaction between CD4 and class II MHC molecules mediates cell adhesion. *Nature* (London) **330:**256–259.

22. **Earl, P. L., R. W. Doms, and B. Moss.** 1990. Oligomeric structure of the HIV-1 envelope glycoprotein. *Proc. Natl. Acad. Sci. USA* **87:**648–652.

23. **Engleman, E. G., C. J. Benike, C. Grumet, and R. L. Evans.** 1981. Activation of human T lymphocyte subsets: helper and suppressor/cytotoxic T cells recognize and respond to distinct histocompatibility antigens. *J. Immunol.* **127:**2124–2129.

24. **Fahey, J. L., J. M. G. Taylor, R. Detels, B. Hofmann, R. Melmed, P. Nishanian, and J. V. Giorgi.** 1990. The prognostic value of cellular and serological markers in infection with human immunodeficiency virus type 1. *N. Engl. J. Med.* **322:**166–172.

25. **Fenouillet, E., B. Clerget-Raslain, J. C. Gluckman, D. Guetard, L. Montagnier, and E. Bahraoui.** 1989. Role of N-linked glycans in the interaction between the envelope glycoprotein of human immunodeficiency virus and its CD4 cellular receptor. *J. Exp. Med.* **169:**807–822.

26. **Fenyö, E. M., L. Morfeldt-Manson, F. Chiodi, B. Lind, A. von Gegerfelt, J. Albert, E. Olausson, and B. Asjö.** 1988. Distinct replicative and cytopathic characteristics of human immunodeficiency virus isolates. *J. Virol.* **62:**4414–4419.

27. **Fisher, R. A., J. M. Bertonis, W. Meier, V. A. Johnson, D. S. Costopoulos, T. Liu, R. Tizard, B. D. Walker, M. S. Hirsh, R. T. Schooley, and R. A. Flavell.** 1988. HIV infection is blocked in vitro by recombinant soluble CD4. *Nature* (London) **331:**76–78.

28. **Folks, T. M., J. Justement, S. R. Mitchell, T. R. Cupps, P. Katz, J. Maples, and A. S. Fauci.** 1987. The T4 epitope is not required for a normal replicative cycles of HIV. *J. Infect. Dis.* **155:**592–593.

29. **Freed, E. O., D. J. Myers, and R. Risser.** 1990. Characterization of the fusion domain of the human immunodeficiency virus type 1 envelope glycoprotein gp41. *Proc. Natl. Acad. Sci. USA* **87**:4650–4654.

30. **Fuller, T. C., J. E. Trevithick, A. A. Fuller, R. B. Colvin, A. B. Cosimi, and P. C. Kung.** 1984. Antigenic polymorphism of the T4 differentiation antigen expressed on human T helper/inducer lymphocytes. *Hum. Immunol.* **9**:89–102.

31. **Gallaher, W. R.** 1987. Detection of a fusion peptide sequence in the transmembrane protein of human immunodeficiency virus. *Cell* **50**:327–328.

32. **Gallo, R. C., S. Z. Salahuddin, M. Popovic, G. M. Shearer, M. Kaplan, B. F. Haynes, T. J. Palker, R. Redfield, J. Oleske, B. Safai, G. White, P. Foster, and P. D. Markham.** 1984. Frequent detection and isolation of cytopathic retroviruses (HTLV-III) from patients with AIDS and at risk for AIDS. *Science* **224**:500–503.

33. **Gomatos, P. J., N. M. Stamatos, H. E. Gendelman, A. Fowler, D. L. Hoover, D. C. Kalter, D. S. Burke, E. C. Tramont, and M. S. Meltzer.** 1990. Relative inefficiency of soluble recombinant CD4 for inhibition of infection of monocyte-tropic HIV in monocytes and T cells. *J. Immunol.* **144**:4183–4188.

34. **Goudsmit, J., C. Debouck, R. H. Meloen, L. Smit, M. Bakker, D. M. Asher, A. V. Wolff, C. J. Gibbs, Jr., and D. C. Gajdusek.** 1988. HIV type 1 neutralization epitope with conserved architecture elicits early type-specific antibodies in experimentally infected chimpanzees. *Proc. Natl. Acad. Sci. USA* **85**:4478–4482.

35. **Guo, H.-G., F. D. Veronese, E. Tschachler, R. Pal, V. S. Kalyanaraman, R. C. Gallo, and M. S. Reitz, Jr.** 1990. Characterization of an HIV-1 point mutant blocked in envelope glycoprotein cleavage. *Virology* **174**:217–224.

36. **Haque, S., K. Saizawa, J. M. Rojo, and C. A. Janeway.** 1987. The influence of valence on the functional activities of monoclonal anti-L3T4 antibodies. *J. Immunol.* **139**:3207–3212.

37. **Hattori, T., A. Koito, K. Takaatsuki, H. Kido, and N. Katunuma.** 1989. Involvement of tryptase-related cellular protease(s) in human immunodeficiency virus type 1 infection. *FEBS Lett.* **248**:48–52.

38. **Ho, D. D., J. C. Kaplan, I. E. Rackauskas, and M. E. Gurney.** 1988. Second conserved domain of gp120 is important for HIV infectivity and antibody neutralization. *Science* **239**:1021–1023.

39. **Hussey, R. E., N. E. Richardson, M. Kowalski, N. R. Brown, H.-C. Chang, R. F. Siliciano, T. Dorfman, B. Walker, J. Sodroski, and E. L. Reinhartz.** 1988. A soluble CD4 protein selectively inhibits HIV replication and syncytium formation. *Nature* (London) **331**:78–81.

40. **Jameson, B. A., P. E. Rao, L. I. Kong, B. H. Hahn, G. M. Shaw, L. E. Hood, and S. B. H. Kent.** 1988. Location and chemical synthesis of a binding site for HIV-1 on the CD4 protein. *Science* **240**:1335–1338.

41. **Janeway, C. A.** 1989. The role of CD4 in T cell activation: accessory molecule or co-receptor? *Immunol. Today* **10**:234–238.

42. **Janeway, C. A., S. Haque, L. A. Smith, and K. Saizawa.** 1987. The role of the murine L3T4 molecule on T cell activation: differential effects of anti-L3T4 on activation by monoclonal antireceptor antibodies. *J. Mol. Cell. Immunol.* **3**:121–131.

43. **Kieny, M., G. Rautmann, D. Schmitt, K. Dott, S. Wain-Hobson, M. Alizon, M. Girard, S. Chamaret, A. Laurent, L. Montagnier, and J. Lecueg.** 1986. AIDS virus env protein expressed from a recombinant vaccinia virus. *Biotechnology* **4**:291–297.

44. **Klatzmann, D., E. Champagne, S. Chamaret, J. Gruest, D. Guetard, T. Hercend, J.-C. Gluckman, and L. Montagnier.** 1984. T-lymphocyte T4 molecule behaves as the receptor for human retrovirus LAV. *Nature* (London) **312**:767–768.

45. **Kowalski, M., J. Potz, L. Basiripour, T. Dorfman, W.C. Goh, E. Terwilliger, A. Dayton, C. Rosen, W. Haseltine, and J. Sodroski.** 1987. Functional regions of the envelope glycoprotein of human immunodeficiency virus type 1. *Science* **237**:1351–1355.

46. Koyanagi, Y., S. Miles, R. T. Mitsuyasu, J. E. Merrill, H. V. Vinters, and I. S. Y. Chen. 1987. Dual infection of the central nervous system by AIDS viruses with distinct cellular tropisms. *Science* **236**:819–822.

47. Kozarsky, K., M. Penman, L. Basiripour, W. Haseltine, J. Sodroski, and M. Kriger. 1989. Glycosylation and processing of the HIV-1 envelope protein. *J. AIDS* **2**:163–169.

48. Krensky, A. M., C. S. Reiss, J. W. Mier, J. L. Strominger, and S. J. Burakoff. 1982. Long-term human cytolytic T-cell lines allospecific for HLA-DR6 antigen are OKT4+. *Proc. Natl. Acad. Sci. USA* **79**:2365–2369.

49. Landau, N. R., M. Warton, and D. R. Littman. 1988. The envelope glycoprotein of the human immunodeficiency virus binds to the immunoglobulin-like domain of CD4. *Nature* (London) **334**:159–162.

50. Lasky, L. A., G. Nakamura, D. H. Smith, C. Fennie, C. Shimasaki, E. Patzer, T. Berman, T. Gregory, and D. J. Capon. 1987. Delineation of a region of the human immunodeficiency virus type 1 gp120 glycoprotein critical for interaction with the CD4 receptor. *Cell* **50**:975–985.

51. Layne, S. P., M. J. Merges, M. Dembo, J. L. Spouge, and P. L. Nara. 1990. HIV requires multiple gp120 molecules for CD4-mediated infection. *Nature* (London) **346**:277–279.

52. Leonard, K. L., W. M. Spellman, L. Riddle, R. J. Harris, J. N. Thomas, and T. J. Gregory. 1990. Assignment of intrachain disulfide bonds and characterization of potential glycosylation sites of the type 1 recombinant human immunodeficiency virus envelope glycoprotein (gp120) expressed in Chinese hamster ovary cells. *J. Biol. Chem.* **265**:10373–10382.

53. Levy, J. A., A. D. Hoffman, S. M. Kramer, J. A. Landis, J. M. Shimabukuro, and L. S. Oshiro. 1984. Isolation of lymphocytopathic retroviruses from San Francisco patients with AIDS. *Science* **225**:840–842.

54. Lifson, J. D., M. B. Feinberg, G. R. Reyes, L. Rabin, B. Banapour, S. Chakrabarti, B. Moss, F. Wong-Staal, K. S. Steimer, and E. G. Engleman. 1986. Induction of CD4-dependent cell fusion by the HTLV-III/LAV envelope glycoprotein. *Nature* (London) **323**:725–728.

55. Lifson, J. D., K. M. Hwang, P. L. Nara, B. Fraser, M. Padgett, N. M. Dunlop, and L. E. Eiden. 1988. Synthetic CD4 peptide derivatives that inhibit HIV infection and cytopathicity. *Science* **241**:712–716.

56. Lifson, J. D., G. R. Reyes, M. S. McGrath, B. S. Stein, and E. G. Engleman. 1986. AIDS retrovirus induced cytopathology: giant cell formation and involvement of CD4 antigen. *Science* **232**:1123–1127.

57. Maddon, P. J., A. G. Dalgleish, J. S. McDougal, P. R. Clapham, R. A. Weiss, and R. Axel. 1986. The T₄ gene encodes the AIDS virus receptor and is expressed in the immune system and the brain. *Cell* **47**:333–348.

58. Maddon, P. J., S. M. Molineaux, D. E. Maddon, K. A. Zimmerman, M. Godfrey, F. W. Alt, L. Chess, and R. Axel. 1987. Structure and expression of the human and mouse T4 genes. *Proc. Natl. Acad. Sci. USA* **84**:9155–9159.

59. Matsushita, S., M. Robert-Guroff, J. Rusche, A. Koito, T. Hattori, H. Hoshino, K. Java-herian, K. Takatsuki, and S. Putney. 1988. Characterization of a HIV neutralizing monoclonal antibody and mapping of the neutralizing epitope. *J. Virol.* **62**:2107–2114.

60. Matthews, T. J., K. J. Weinhold, H. K. Lyerly, A. J. Langlois, H. Wigzell, and D. P. Bolognesi. 1987. Interaction between the human T-cell lymphotropic virus type III_B envelope glycoprotein gp120 and the surface antigen CD4: role of carbohydrate in binding and cell fusion. *Proc. Natl. Acad. Sci. USA* **84**:5424–5428.

61. Mayer, W. E., M. Jonker, D. Klein, G. van Seventer, and J. Klein. 1988. Nucleotide sequences of chimpanzee MHC class I alleles: evidence for trans-species mode of evolution. *EMBO J.* **7**:2765–2774.

62. McClure, M. O., M. Marsh, and R. A. Weiss. 1988. Human immunodeficiency virus infection of CD4-bearing cells occurs by a pH-independent mechanism. *EMBO J.* **7**:513–518.

63. **Meuer, S. C., S. F. Schlossman, and E. L. Reinherz.** 1982. Clonal analysis of human cytotoxic T lymphocytes. T4+ and T8+ effector T cells recognize products of different major histocompatibility complex regions. *Proc. Natl. Acad. Sci. USA* **79**:4395–4399.

64. **Mizouchi, T., M. W. Spellman, M. Larkin, J. Soloman, L. J. Basa, and T. Feizi.** 1988. Carbohydrate structures of the HIV recombinant envelope glycoprotein gp120 produced in Chinese hamster ovary cells. *Biochem. J.* **254**:599–603.

65. **Mizukami, T., T. R. Fuerst, E. A. Berger, and B. Moss.** 1988. Binding region for human immunodeficiency virus (HIV) and epitopes of HIV-blocking monoclonal antibodies of the CD4 molecule defined by site-directed mutagenesis. *Proc. Natl. Acad. Sci. USA* **85**:9273–9277.

66. **Modrow, S., B. H. Hahn, G. M. Shaw, R. C. Gallo, F. Wong-Staal, and H. Wolf.** 1987. Computer-assisted analysis of envelope protein sequences of seven human immunodeficiency virus isolates: prediction of antigenic epitopes in conserved and variable regions. *J. Virol.* **61**:570–578.

67. **Moscicki, R. A., E. P. Amento, S. M. Krane, J. T. Kurnick, and R. B. Colvin.** 1983. Modulation of surface antigens of a human monocyte cell line, U937, during incubation with T lymphocyte-conditioned medium: detection of T4 antigen and its presence on normal blood monocytes. *J. Immunol.* **131**:743–748.

68. **Muesing, M. A., D. H. Smith, C. D. Cabradilla, C. V. Benton, L. A. Lasky, and D. J. Capon.** 1985. Nucleic acid structure and expression of the human AIDS/lymphadenopathy retrovirus. *Nature* (London) **313**:450–458.

69. **Mustelin, T., and A. Altman.** 1989. Do CD4 and CD8 control T-cell activation via a specific tyrosine protein kinase? *Immunol. Today* **10**:189–192.

70. **Myers, G., A.B. Rabson, S. F. Josephs, T. F. Smith, J.A. Berzofsky, and F. Wong-Staal (ed.).** 1990. *Human Retroviruses and AIDS 1990.* Los Alamos National Laboratory, Los Alamos, N. Mex.

71. **O'Brien, W. A., Y. Koyanagi, A. Namazie, J.-Q. Zhao, A. Diagne, K. Idler, J. A. Zack, and I. S. Y. Chen.** 1990. HIV-1 tropism for mononuclear phagocytes can be determined by regions of gp120 which do not include the CD4 binding domain. *Nature* (London) **348**:69–73.

72. **Palker, T. J., M. E. Clark, A. J. Langlois, J. T. Matthews, K. J. Weinhold, R. R. Randall, D. P. Bolognesi, and B. F. Haynes.** 1988. Type-specific neutralization of HIV with antibodies to env-encoded synthetic peptides. *Proc. Natl. Acad. Sci. USA* **85**:1932–1936.

73. **Peterson, A., and B. Seed.** 1988. Genetic analysis of monoclonal antibody and HIV binding sites on the human lymphocyte antigen, CD4. *Cell* **54**:65–72.

74. **Pinter, A., W. J. Honnen, S. A. Tilley, C. Bona, H. Zaghouani, M. K. Gorny, and S. Zolla-Pazner.** 1989. Oligomeric structure of gp41, the transmembrane protein of HIV-1. *J. Virol.* **63**:2674–2679.

75. **Popovic, M., M. G. Sarngadharan, E. Read, and R. C. Gallo.** 1984. Detection, isolation, and continuous production of cytopathic retroviruses (HTLV-III) from patients with AIDS and pre-AIDS. *Science* **224**:497–500.

76. **Qureshi, N. M., D. H. Coy, R. F. Garry, and L. A. Henderson.** 1990. Characterization of a putative cellular receptor for HIV-1 transmembrane protein using synthetic peptides. *AIDS Res. Hum. Retroviruses* **4**:553–558.

77. **Ratner, L., W. Haseltine, R. Patarca, K. J. Livak, B. Starcich, S. F. Josephs, E. R. Doran, J. A. Rafalski, E. A. Whitehorn, K. Baumeister, L. Ivanoff, S. R. Petteway, Jr., M. L. Pearson, J. A. Lautenberger, T. S. Papas, J. Ghrayeb, N. T. Chang, R. C. Gallo, and F. Wong-Staal.** 1985. Complete nucleotide sequence of the AIDS virus, HTLV-III. *Nature* (London) **313**:277–284.

78. **Reinherz, E. L., P. C. King, G. Goldstein, and S. F. Schlossman.** 1979. Separation of func-

tional subsets of human T cells by a monoclonal antibody. *Proc. Natl. Acad. Sci. USA* **76**:4061–4065.

79. **Robey, W. G., B. Safai, S. Oroszlan, L. O. Arthur, M. A. Gonda, R. C. Gallo, and P. J. Fischinger.** 1985. Characterization of envelope and core structural gene products of HTLV-III with sera from AIDS patients. *Science* **228**:593–595.

80. **Rojo, J. M., and C. A. Janeway.** 1988. The biologic activity of anti-T cell receptor V region antibodies is determined by the epitope recognized. *J. Immunol.* **140**:1081–1088.

81. **Rojo, J. M., K. Saizawa, and C. A. Janeway.** 1989. Physical association of CD4 and the T-cell receptor can be induced by anti-T cell receptor antibodies. *Proc. Natl. Acad. Sci. USA* **86**:3311–3315.

82. **Rudd, C. E., J. M. Trevillyan, J. D. Dasgupta, L. L. Wong, and S. F. Schlossman.** 1988. The CD4 receptor is complexed in detergent lysates to a protein-tyrosine kinase (pp58) from human T lymphocytes. *Proc. Natl. Acad. Sci. USA* **85**:5190–5194.

83. **Rusche, J. R., K. Javaherian, C. McDanal, J. Petro, D. L. Lynn, R. Grimaila, A. Langlois, R. C. Gallo, L. A. Arthur, P. J. Fischinger, D. P. Bolognesi, S. D. Putney, and T. J. Matthews.** 1988. Antibodies that inhibit fusion of HIV-infected cells bind a 24 amino-acid sequence of the viral envelope, gp120. *Proc. Natl. Acad. Sci. USA* **85**:3198–3202.

84. **Saizawa, K., J. M. Rojo, and C. A. Janeway.** 1987. Evidence for a physical association of CD4 and the CD3:α:β T-cell receptor. *Nature* (London) **328**:260–263.

85. **Sakai, K., S. Dewhurst, X. Ma, and D. J. Volsky.** 1988. Differences in cytopathogenicity and host range among infectious molecular clones of human immunodeficiency virus type 1 simultaneously isolated from an individual. *J. Virol.* **62**:4078–4085.

86. **Sanchez-Pescador, R., M. D. Power, P. J. Barr, K. S. Steimer, M. M. Stempien, S. L. Brown-Shimer, W. W. Gee, A. Renard, A. Randolph, J. A. Levy, D. Dina, and P. A. Luciw.** 1985. Nucleotide sequence and expression of an AIDS-associated retrovirus (ARV-2). *Science* **227**:484–492.

87. **Sattentau, Q. J., J. Arthos, K. Deen, N. Hanna, D. Healey, P. C. L. Beverley, R. Sweet, and A. Truneh.** 1989. Structural analysis of the human immunodeficiency virus-binding domain of CD4. Epitope mapping with site-directed mutants and anti-idiotypes. *J. Exp. Med.* **170**:1319–1334.

88. **Shapira-Nahor, O., J. Golding, L. K. Vujcic, S. Resto-Ruiz, R. L. Fields, and F. A. Robey.** 1990. CD4-derived synthetic peptide blocks the binding of HIV-1 gp120 to CD4-bearing cells and prevents HIV-1 infection. *Cell. Immunol.* **128**:101–117.

89. **Shaw, A. S., K. E. Amrein, C. Hammond, D. F. Stern, B. M. Sefton, and J. K. Rose.** 1989. The lck tyrosine protein kinase interacts with the cytoplasmic tail of the CD4 glycoprotein through its unique amino terminal domain. *Cell* **59**:627–636.

90. **Sodroski, J., W. C. Goh, C. Rosen, K. Campbell, and W. Haseltine.** 1986. Role of the HTLV-III/LAV envelope in syncytium formation and cytopathogenicity. *Nature* (London) **322**:470–474.

91. **Stein, B. S., S. D. Gowda, J. D. Lifson, R. C. Penhallow, K. G. Bensch, and E. G. Engleman.** 1987. pH-independent HIV entry into CD4-positive T cells via virus envelope fusion to the plasma membrane. *Cell* **49**:659–668.

92. **Stephens, P. E., G. Clements, G. T. Yarranton, and J. Moore.** 1990. A chink in HIV's armour? *Nature* (London) **343**:219.

93. **Sun, N.-C., D. D. Ho, C. R. Y. Sun, R.-S. Liou, W. Gordon, M. S. C. Fung, X.-L. Li, R. C. Ting, T.-H. Lee, N. T. Chang, and T.-W. Chang.** 1989. Generation and characterization of monoclonal antibodies to the putative CD4-binding domain of human immunodeficiency virus type 1 gp120. *J. Virol.* **63**:3579–3585.

94. **Syu, W.-J., J.-H. Huang, M. Essex, and T.-H. Lee.** 1990. The N-terminal region of the HIV envelope glycoprotein gp120 contains potential binding sites for CD4. *Proc. Natl. Acad. Sci. USA* **87**:3695–3699.

95. Tersmette, M., R. E. Y. de Goede, B. J. M. Al, I. N. Winkel, R. A. Gruters, H. T. Cuypers, H. G. Huisman, and F. Miedema. 1988. Differential syncytium-inducing capacity of human immunodeficiency virus isolates: frequent detection of syncytium-inducing isolates in patients with acquired immunodeficiency syndrome (AIDS) and AIDS-related complex. *J. Virol.* **62:**2026–2032.

96. Tersmette, M., R. A. Gruters, F. de Wolf, R. E. Y. de Goede, J. M. A. Lange, P. T. A. Schellekens, J. Goudsmit, H. G. Huisman, and F. Miedema. 1989. Evidence for a role of virulent human immunodeficiency virus (HIV) variants in the pathogenesis of acquired immunodeficiency syndrome: studies on sequential HIV isolates. *J. Virol.* **63:**2118–2125.

97. Tourvielle, B., S. D. Gorman, E. H. Field, T. Hunkapiller, and J. R. Parnes. 1986. Isolation and sequence of L3T4 complementary DNA clones: expression in T cells and brain. *Science* **234:**610–614.

98. Tschachler, E., H. Buchow, R. C. Gallo, and M. S. Reitz, Jr. 1990. Functional contribution of cysteine residues to the human immunodeficiency virus type 1 envelope. *J. Virol.* **64:**2250–2259.

99. Veillette, A., M. A. Bookman, E. M. Horak, and J. B. Bolen. 1988. The CD4 and CD8 T cell surface antigens are associated with the internal membrane tyrosine-protein kinase p56[lck]. *Cell* **55:**301–308.

100. von Briesen, H., W. B. Becker, K. Henco, E. B. Helm, H. R. Gelderbom, H. D. Brede, and H. Rubsamen-Waigmann. 1987. Isolation frequency and growth properties of HIV-variants: multiple simultaneous variants in a patient demonstrated by molecular cloning. *J. Med. Virol.* **23:**51–66.

101. Wain-Hobson, S., P. Sonigo, O. Danos, S. Cole, and M. Alizon. 1985. Nucleotide sequence of the AIDS virus, LAV. *Cell* **40:**9–17.

102. Willey, R. L., D. H. Smith, L. A. Lasky, T. S. Theodore, P. L. Earl, B. Moss, D. J. Capon, and M. A. Martin. 1988. In vitro mutagenesis identifies a region within the envelope gene of the human immunodeficiency virus that is critical for infectivity. *J. Virol.* **62:**139–147.

103. Wood, G. S., N. L. Warner, and R. A. Warnke. 1983. Anti-Leu-3/T4 antibodies react with cells of monocyte/macrophage and Langerhans lineage. *J. Immunol.* **131:**212–216.

Viruses That Affect the Immune System
Edited by Hung Y. Fan et al.
© 1991 American Society for Microbiology, Washington, DC 20005

Chapter 6

CD4: Function, Structure, and Interactions with the Human Immunodeficiency Virus Type 1 Envelope Protein gp120

David C. Diamond, Michael R. Bowman, Kurtis D. MacFerrin, Stuart L. Schreiber, and Steven J. Burakoff

CD4 is a cell surface glycoprotein found primarily on class II major histo-compatibility complex (MHC) restricted T lymphocytes, and it is a receptor whose ligand is a monomorphic determinant on class II MHC antigens (12, 31). In addition to its role in antigen recognition by and activation of T cells (see reference 4 and below), CD4 is also a receptor for human immunode-ficiency virus type 1 (HIV-1) (8, 18, 25, 27), the etiologic agent of acquired immunodeficiency syndrome (AIDS). Here we report our studies concerning the function and structure of CD4 and how it interacts with gp120, the en-velope protein of HIV-1.

We have made extensive use of a model system consisting of a murine T-cell hybridoma whose antigen-specific T-cell receptor recognizes the human class II MHC antigen HLA-DR. Spleen cells from mice immunized with an HLA-DR+ human cell line (JY) were fused with the murine thymoma line BW5147. A hybridoma line was isolated from this fusion which produces interleukin-2 (IL-2) when stimulated with HLA-DR+ cells. This hybridoma was then infected with a retroviral vector carrying a cDNA encoding human

David C. Diamond • Baxter Hyland Division, Duarte, California 91010. **Michael R. Bow-man** • BASF Bioresearch Corp., Cambridge, Massachusetts 02139. **Kurtis D. MacFerrin and Stuart L. Schreiber** • Department of Chemistry, Harvard University, Cam-bridge, Massachusetts 02138. **Steven J. Burakoff** • Division of Pediatric Oncology, Dana Farber Cancer Institute, and Department of Pediatrics, Harvard Medical School, Boston, Mas-sachusetts 02115.

CD4 and the selectable marker *neo,* thereby conferring expression of CD4 and resistance to the antibiotic G418 (37).

Expression of CD4 in this hybridoma results in enhanced IL-2 production as compared with the parent line or hybridomas expressing only the *neo* gene. CD4 expression results in 5- to 15-fold greater IL-2 production when stimulated by HLA-DR$^+$ cells compared with the CD4$^-$ hybridomas. This enhanced response to antigen is inhibitable by anti-CD4 monoclonal antibodies (MAbs), thus demonstrating that the enhanced response is due to CD4 expression. CD4-mediated conjugate formation can also be demonstrated between the CD4$^+$ hybridoma and HLA-DR$^+$ liposomes or cells (37). Conjugate formation by this hybridoma can also be blocked by several anti-CD4 MAbs (31).

The viral attachment protein of HIV-1 is its envelope protein, gp120, which binds to CD4 with high affinity (22, 27). Several of the anti-CD4 MAbs which block conjugate formation and IL-2 production by the CD4$^+$ hybridoma (31, 37) also block gp120 binding and infection by HIV-1 (18, 22, 27, 32). This raised the possibility that gp120 and class II MHC might share overlapping binding sites on CD4 and that gp120 might also block CD4-dependent conjugate formation and IL-2 production. This offered a potential explanation of the immune suppression seen in AIDS patients even before depletion of CD4$^+$ T cells (13, 20).

When gp120 is incubated with the CD4$^+$ hybridoma and HLA-DR$^+$ cells, IL-2 production is inhibited (Fig. 1) in a concentration-dependent manner and, based on the published K_d of 4.18 nM (22), to a degree consistent with the degree of occupancy of CD4 by gp120 (11). Complete inhibition of the CD4-dependent IL-2 production is observed with near-saturating concentrations of gp120. As expected from the competitive mechanism envisioned for this inhibition, increasing the number of HLA-DR$^+$ cells used to stimulate the hybridoma led to reversal of the inhibition. That the effect of gp120 is CD4 mediated is further demonstrated by its inability to inhibit IL-2 production by a CD4$^-$ CD2$^+$ hybridoma derived from the same parent line. The gp120 concentrations used (0.45 to 5.4 μg/ml) are greater than would be expected systemically in the blood stream but could conceivably be attained locally at a focus of infection; thus this phenomenon may be physiologically relevant.

The importance of the cytoplasmic domain of CD4 has also been investigated. The cytoplasmic domain of CD4 associates with the lymphocyte-specific tyrosine kinase Lck (42) and also contains three serines which become phosphorylated upon activation with phorbol myristic acetate or antigen (4, 34). Thus it is possible that the cytoplasmic domain of CD4 may transduce a biochemical signal crucial to IL-2 production or its inhibition by gp120. While mutation of the three serines to alanine or deletion of 31 (including

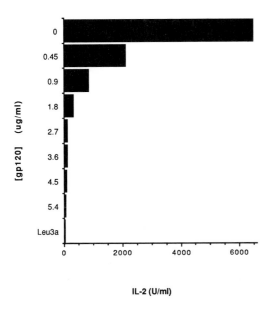

Figure 1. Inhibition of CD4-dependent IL-2 production by gp120. The HLA-DR-responsive hybridoma 16CD4-9 was stimulated with the HLA-DR⁺ cell line Daudi in the presence of various concentrations of gp120 or 400 ng of the anti-CD4 MAb Leu3a per ml. IL-2 production was assayed as the ability to support proliferation ([³H]thymidine incorporation) of the CTLL20 cell line. 100 U/ml was defined as the half-maximal proliferation, supported by a standard rat concanavalin A supernatant. The background from unstimulated 16CD4-9 was <10 U/ml. See reference 11.

the three serines) of the 38 cytoplasmic amino acids does disrupt the ability of these CD4 molecules to modulate in response to phorbol myristic acetate (26, 34, 35), they are nonetheless still able to mediate CD4-dependent IL-2 production in response to HLA-DR⁺ cells (36; unpublished observations), although the degree of enhancement of IL-2 production is reduced for the deletion mutant (36; unpublished observations). Similarly, deletion of the cytoplasmic domain does not disrupt the ability of gp120 to inhibit IL-2 production (11). Thus the cytoplasmic domain of CD4 does not appear to be crucial to the inhibition seen with gp120 when the hybridomas are stimulated by antigen. In contrast, when the membrane-spanning region of CD4 is deleted as well and CD4 is instead anchored in the membrane by glycosyl phosphatidylinositol (GPI), CD4-dependent IL-2 production is not observed (unpublished observations). This molecule (CD4PI), however, does retain normal conjugate formation and gp120 binding activity, as does the cytoplasmic domain deletion mutant (unpublished observations).

It has been shown that alterations of the cytoplasmic and membrane-spanning regions and deletion of the cytoplasmic domain of CD4 disrupt modulation of CD4 but fail to disrupt the ability of CD4 to mediate infection by HIV-1 (3, 25). Since similar mutants retain the CD4 function of enhancing IL-2 production (36; unpublished observations), we were interested to see if the CD4PI molecule, which did not retain this CD4 function, has also lost its ability to mediate productive infection by HIV-1. To this end the retroviral vector carrying the CD4PI gene construct was used to infect the human pre-T-cell leukemia line HSB2. Treatment with phosphatidylinositol-specific phospholipase C released up to 70% of the protein expressed from the CD4PI gene construct in HSB2 (10). Incomplete removal by phosphatidylinositol-specific phospholipase C of GPI-anchored proteins is common, possibly due to acylation of the inositol ring, among other factors (24). To demonstrate that indeed none of the protein synthesized from the CD4PI gene construct becomes anchored in the plasma membrane by a peptide sequence, we have made use of L cells, which fail to attach GPI anchors and instead secrete proteins normally anchored by GPI (39). When CD4PI is expressed in L cells, soluble CD4 is detectable in the culture supernatant, but no CD4 is detectable on the cell surface by flow cytometry. When CD4PI-expressing HSB2 cells were tested it was found that CD4PI is at least as efficient as wild-type CD4 (expressed in HSB2) in mediating HIV-1 infection (10). Thus it can be concluded that CD4 function is not required for productive infection by HIV-1. These data also demonstrate that neither the membrane-spanning nor the cytoplasmic domain of CD4 is required as structural elements for viral receptor function. Furthermore, these results suggest the potential of GPI-anchored proteins to serve as viral receptors.

The first immunoglobulinlike domain of CD4 has been shown to contain the primary gp120 binding determinant and amino acids 42 through 55, identified by homolog scanning mutational analysis as being of particular importance (1,6). Essential determinants for class II MHC binding appear to be more widely distributed but overlap with the gp120 binding site in the region of amino acids 48 to 51 (7, 19). However, the roles of individual residues are obscured in these studies since blocks of amino acids have been substituted. Also the homologous nature of the substituted blocks of sequence, which could carry compensating differences, may mask regions of importance. This seems to have happened in the study by Lamarre et al. (19), where four amino acids between amino acids 48 and 52 were exchanged to no effect, while in contrast, when Clayton et al. (7) exchanged three of these residues, both gp120 and class II MHC binding were lost.

We have examined a series of anti-CD4 MAb epitope loss mutations, concentrating on point mutations, in the first (N-terminal) immunoglobulin-like domain of CD4. We have assessed the ability of these mutants to bind gp120 and class II MHC, as well as their ability to mediate enhanced IL-2 production in response to antigen. The mutants were originally selected on the basis of their loss of individual anti-CD4 MAb epitopes (30). In addition we have constructed a three-dimensional model of CD4 structure in this region to aid in interpretation of the data.

The model of CD4 was constructed by a series of sequence alignments and use of computerized structure building and energy minimization routines (see reference 5 for a complete description). First, the CD4 sequences from several species were aligned in order to emphasize conserved structural elements. These were then aligned as a set with an aligned set of immunoglobulin variable regions. The REI light-chain dimer, chosen because it has the greatest amino acid sequence identity with CD4, was then used as a framework on which to build the CD4 model. Three areas of insertion or deletion between the two sequences requiring some reconstruction of the backbone were identified. One of these, near the turn in the C-C' β-sheet, required substantial changes in the geometry of the adjacent sequences. To maintain the conserved β-turn connecting the two strands, it was necessary to lengthen the C strand relative to the C' strand, thereby giving the C-C' β-sheet a twist of the opposite handedness from that in REI. Next, the nonidentical side chains from CD4 were substituted into the rest of the structure, and finally the entire structure was subjected to energy minimization. In contrast, previous models have tended to simply assign CD4 side chains to the REI backbone, allowing insertions and deletions with little regard for conserved structural elements.

Five CD4 epitope loss mutants, a deletion of amino acids 42 through 49, and four point mutants, Q40P (denoting glutamine 40 changed to proline), F43L, G47R, and P48S, have been characterized (5). These mutant CD4 molecules were each expressed in the HLA-DR-responsive T-cell hybridoma discussed above and tested for MAb binding with a panel of anti-CD4 MAbs (T4, MT151, Leu3a, and the OKT4 series) and for gp120 binding, CD4-class II MHC-mediated conjugate formation, and enhanced IL-2 production (Table 1).

The mutants exhibiting the most widespread effects were the 42–49 deletion mutant and Q40P, which disrupt binding by most of the panel of MAbs and gp120, conjugate formation, and, consequently, enhanced IL-2 production. From our model the deletion would be expected to remove an entire

Table 1. Effects of Mutations in the V1 Domain of CD4[a]

Mutation	MAb binding	gp120 binding	Class II MHC binding	IL-2 enhancement
CD4 wild type	+	+	+	+
42–49 amino-acid deletion	–	–	–	–
Q40P	–	–	–	–
F43L	+	–	–	–
G47R	+	–	–	–
P48S	+	+	+	+

[a]For MAb binding, + indicates wild-type levels of binding for at least 9 of 10 tested MAbs (see text); + indicates a level comparable to the CD4wt+ hybridoma; – indicates a level comparable to the CD4- hybridoma. See reference 5.

β-strand connecting distant parts of the protein, while the substitution of a proline next to G41 could very well induce a β-turn in the middle of a β-strand. Clearly both of these changes would be disruptive to the overall structure of CD4 and as such provide little information about the direct involvement of these residues in gp120 and class II MHC binding. At the other extreme, P48S has no major effect on any of the tested parameters. This position is predicted to be in a loop, and thus changes there cannot readily exert major structural effects. Additionally, proline and serine can often fulfill similar structural roles. Still, the prolyl side chain does not appear to be a critical determinant of either gp120 or class II MHC binding.

The remaining two mutations are both located on the exposed (C') side of the C-C' β-sheet, a highly accessible region of CD4. Both F43L and G47R disrupt gp120 and class II MHC binding without inducing major structural changes detectable by MAb binding. While the substitution of the much bulkier arginine residue for G47 will restrict the possible conformations at this position, it is also a radical change, introducing both charge and bulk, which could easily interfere with binding interactions. The F43L mutation would not be expected to significantly affect the peptide backbone conformation. Therefore, the most likely explanation for the loss of binding is modification of a side chain contact.

Based on the importance of Phe-43 to gp120 binding, a phenylalanine-containing dipeptide, termed CPF, blocked at both termini was included in a screen for gp120 binding inhibitors and found to be active. Using a flow cytometric assay of gp120 binding to cellular CD4, we have begun to characterize the activity of this compound and several congeners (14). Inhibition of gp120 binding is detected as a reduction in mean fluorescence. The activity of the CPF appears to be derived from its combination of side chains and blocking groups and their relative positioning, though it is not sensitive to the stereochemistry of its backbone. All four combinations of L- and D-amino acids in the CPF have been synthesized and found to have very similar activities. Of the four stereoisomers, the D-amino acid-containing analog is the most active. This is a fortunate occurrence in terms of drug development potential since it should be resistant to proteases and thus have greater stability. Substitution of alanine for the phenylalanine residue greatly diminished its activity. The blocked single amino acids are inactive. We are testing a variety of other congeners to further define the basis of the activity.

This compound appears to interact directly with gp120, unlike another small molecule inhibitor of gp120-CD4 binding, aurintricarboxylic acid, which alters CD4 conformation (33). If, after preincubation, the gp120-CPF mixture is diluted and the gp120 is reconcentrated by ultrafiltration, the gp120 is still inhibited from binding to CD4. In fact, several cycles of dilution and reconcentration fail to dissociate the CPF from gp120, implying a much better affinity constant than indicated by simple dilution of the CPF in the binding inhibition assay. The CPF is not directly soluble in aqueous solution but must be first dissolved in dimethyl sulfoxide and then diluted into solution. We suspect that there may be a partitioning effect between dimethyl sulfoxide and aqueously solvated CPF and that the latter is the active species. We are currently trying to design a more soluble, and thus perhaps more active, congener.

The physiologic ligands of CD4 are the class II MHC proteins (4, 12, 31). As discussed above, this interaction has been demonstrated to promote conjugate formation in vitro between human CD4[+] murine T-cell hybridomas and class II MHC[+] cells (31), and in the same system CD4 expression leads to increased responsiveness to antigenic stimulation as measured by IL-2 production (37). By both these parameters, gp120 inhibits CD4 function (11, 31). The inhibition of conjugate formation by gp120 can be reversed by the CPF but not by its less active analogs, either after incubation with gp120 or with simultaneous addition of all components (14). In fact, addition of the CPF restores conjugate formation where conjugates have not formed due to prior gp120 inhibition. Of equal importance, these compounds have no effect on conjugate formation directly. Thus the CPF would not be expected to interfere with CD4 function. To the contrary, it should reverse any inhibition due to gp120.

Ultimately we wanted to determine if the activity of the CPF could be detected in more biological, as opposed to biochemical, assay systems. Experiments also show that the CPF can also inhibit in vitro infection with HIV-1. Virus preincubated with the CPF inhibited infection of H9 cells, assayed as p24 core antigen production, while a congener which did not block gp120 binding did not inhibit infectivity (14). In CD4-dependent production of IL-2, we have found that gp120, pretreated with the CPF and then diluted and reconcentrated by ultrafiltration, is incapable of inhibiting IL-2 production (14). Thus, the CPF appears to prevent gp120 inhibition of MHC class II[+] stimulation of the CD4[+] hybridoma.

Inhibition of viral attachment is an attractive antiviral strategy. It acts at the earliest stages in the cycle of infection against a target, the viral attachment protein/site, that must be maintained for infectivity. Resistance to enzyme inhibitors might be more readily established than resistance to receptor-ligand blockade. Indeed, strains of HIV-1 resistant to azidothymidine, the only anti-HIV drug in widespread use, have already begun to be detected, although their clinical importance has yet to be evaluated (21). To date the application of this strategy has concentrated on receptor blockade using CD4 analogs, either engineered proteins (9, 15, 16, 38, 40, 41) and synthetic peptides (17, 23, 29) or polysulfated anions, e.g., dextran sulfate (2, 28). Both classes of agents would probably require administration by injection to be effective. Additionally, the mode of action of the polysulfated anions is relatively nonspecific (2, 28, 33). The potential of a drug developed from the CPFs could be great. Its small size and incorporation of D-amino acids might lead to easy (oral) administration and extended biologic half-life. Clinically, both disruption of the spread of the infection within the individual and restoration of normal CD4 function may be anticipated. Based on our increasing knowledge of CD4 structure and function and our understanding of the CPF's activity, we are continuing to work toward this goal.

REFERENCES

1. Arthos, J., K. C. Deen, M. A. Chaikin, J. A. Fornwald, G. Sathe, Q. J. Sattentau, P. R. Clapham, R. A. Weiss, J. S. McDougal, C. Pietropaolo, R. Axel, A. Truneh, P. J. Maddon, and R. W. Sweet. 1989. Identification of the residues in human CD4 critical for the binding of HIV. *Cell* 57:469–481.
2. Baba, M., R. Pauwels, J. Balzarini, J. Arnout, J. Desmyter, and E. de Clercq. 1988. Mechanism of inhibitory effect of dextran sulfate and heparin on replication of human immunodeficiency virus in vitro. *Proc. Natl. Acad. Sci. USA* 85:6132–6136.
3. Bedinger, P., A. Moriarty, R. C. von Borstel, N. J. Donovan, K. S. Steimer, and D. R. Littman. 1988. Internalization of the human immunodeficiency virus does not require the cytoplasmic domain of CD4. *Nature* (London) 334:162–165.
4. Bierer, B. E., B. P. Sleckman, S. E. Ratnofsky, and S. J. Burakoff. 1989. The biologic roles of CD2, CD4, and CD8 in T cell activation. *Annu. Rev. Immunol.* 7:579–599.
5. Bowman, M. R., K. D. MacFerrin, S. L. Schreiber, and S. J. Burakoff. 1990. Identification and structural analysis of residues in the V1 region of CD4 involved in the interaction with HIV gp120 and class II MHC. *Proc. Natl. Acad. Sci. USA* 87:9052–9056.
6. Clayton, L. K., R. E. Hussey, R. Steinbrich, H. Ramachandran, Y. Husain, and E. L. Reinherz. 1988. Substitution of murine for human CD4 residues identifies amino acids critical for HIV-gp120 binding. *Nature* (London) 335:363–366.
7. Clayton, L. K., M. Sieh, D. A. Pious, and E. L. Reinherz. 1989. Identification of human CD4 residues affecting class II MHC versus HIV-1 gp120 binding. *Nature* (London) 339:548–551.
8. Dalgleish, A. G., P. C. L. Beverly, P. R. Clapham, D. H. Crawford, M. F. Greaves, and R. A. Weiss. 1984. The CD4 (T4) antigen is an essential component of the receptor for the AIDS retrovirus. *Nature* (London) 312:763–766.
9. Deen, K. C., J. S. McDougal, R. Inacker, G. Folena-Wasserman, J. Arthos, J. Rosenberg, P. J. Maddon, R. Axel, and R. W. Sweet. 1988. A soluble form of CD4 (T4) protein inhibits AIDS virus infection. *Nature* (London) 331:82–86.
10. Diamond, D. C., R. Finberg, S. Chaudhuri, B. P. Sleckman, and S. J. Burakoff. 1990. Human immunodeficiency virus infection is efficiently mediated by a glycolipid anchored form of CD4. *Proc. Natl. Acad. Sci. USA* 87:5001–5005.
11. Diamond, D. C., B. P. Sleckman, T. Gregory, L. A. Lasky, J. L. Greenstein, and S. J. Burakoff. 1988. Inhibition of CD4⁺ T cell function by the HIV envelope protein, gp120. *J. Immunol.* 141:3715–3717.
12. Doyle, C., and J. L. Strominger. 1987. Interaction between CD4 and class II MHC molecules mediates cell adhesion. *Nature* (London) 330:256–258.
13. Fauci, A. S. 1988. The human immunodeficiency virus: infectivity and mechanism of pathogenesis. *Science* 239:617–622.
14. Finberg, R. W., D. C. Diamond, D. B. Mitchell, Y. Rosenstein, G. Soman, T. C. Norman, S. L. Schreiber, and S. J. Burakoff. 1990. Prevention of HIV-1 infection and preservation of CD4 function by the binding of CPFs to gp120. *Science* 249:287–291.

15. Fisher, R. A., J. M. Bertonis, W. Meier, V. A. Johnson, D. S. Costopoulos, T. Liu, R. Tizard, B. D. Walker, M. S. Hirsch, R. T. Schooley, and R. A. Flavell. 1988. HIV infection is blocked in vitro by recombinant soluble CD4. *Nature* (London) 331:76–78.

16. Hussey, R. E., N. E. Richardson, M. Kowalski, N. R. Brown, H.-C. Chang, R. F. Siliciano, T. Dorfman, B. Walker, J. Sodroski, and E. L. Reinherz. 1988. A soluble CD4 protein selectively inhibits HIV replication and syncytium formation. *Nature* (London) 331:78–81.

17. Jameson, B. A., P. E. Rao, L. I. Kong, B. H. Hahn, G. M. Shaw, L. E. Hood, and S. B. H. Kent. 1988. Location and chemical synthesis of a binding site for HIV-1 on the CD4 protein. *Science* 240:1335–1338.

18. Klatzman, D., E. Champagne, S. Chamaret, J. Gruest, D. Guetard, T. Hercend, J.-C. Gluckman, and L. Montagnier. 1984. T-lymphocyte T4 molecule behaves as the receptor for human retrovirus LAV. *Nature* (London) 312:767–768.

19. Lamarre, D., A. Ashkenazi, S. Fleury, D. H. Smith, R.-P. Sekaly, and D. J. Capon. 1989. The MHC-binding and gp120-binding functions of CD4 are separable. *Science* 245:743–746.

20. Lane, H. C., and A. S. Fauci. 1985. Immunologic abnormalities in the acquired immunodeficiency syndrome. *Annu. Rev. Immunol.* 3:477–500.

21. Larder, B. A., G. Darby, and D. D. Richman. 1989. HIV with reduced sensitivity to Zidovudine (AZT) isolated during prolonged therapy. *Science* 243:1731–1734.

22. Lasky, L. A., G. Nakamura, D. H. Smith, C. Fennie, C. Shimasaki, E. Patzer, P. Berman, T. Gregory, and D. J. Capon. 1987. Delineation of a region of the human immunodeficiency virus type 1 gp120 glycoprotein critical for interaction with the CD4 receptor. *Cell* 50:975–985.

23. Lifson, J. D., K. M. Hwang, P. L. Nara, B. Fraser, M. Padgett, N. M. Dunlop, and L. E. Eiden. 1988. Synthetic CD4 peptide derivatives that inhibit HIV infection and cytopathicity. *Science* 241:712–716.

24. Low, M. G. 1989. Glycosyl-phosphatidylinositol: a versatile anchor for cell surface proteins. *FASEB J.* 3:1600–1608.

25. Maddon, P. J., A. G. Dalgleish, J. S. McDougal, P. R. Clapham, R. A. Weiss, and R. Axel. 1986. The T4 gene encodes the AIDS virus receptor and is expressed in the immune system and the brain. *Cell* 47:333–348.

26. Maddon, P. J., J. S. McDougal, P. R. Clapham, A. G. Dalgleish, S. Jamal, R. A. Weiss, and R. Axel. 1988. HIV infection does not require endocytosis of its receptor, CD4. *Cell* 54:865–874.

27. McDougal, J. S., M. S. Kennedy, J. M. Sligh, S. P. Cort, A. Mawle, and J. K. A. Nicholson. 1986. Binding of HTLV-III/LAV to T4+ T cells by a complex of the 110K viral protein and the T4 molecule. *Science* 231:382–385.

28. Mitsuya, H., D. J. Looney, S. Kuno, R. Ueno, F. Wong-Staal, and S. Broder. 1988. Dextran sulfate suppression of viruses in the HIV family: inhibition of virion binding to CD4+ cells. *Science* 240:646–649.

29. Nara, P. L., K. M. Hwang, D. M. Rausch, J. D. Lifson, and L. E. Eiden. 1989. CD4 antigen-based antireceptor peptides inhibit infectivity of human immunodeficiency virus in vitro at multiple stages of the viral life cycle. *Proc. Natl. Acad. Sci. USA* 86:7139–7143.

30. Peterson, A., and B. Seed. 1988. Genetic analysis of monoclonal antibody and HIV-binding sites on the human lymphocyte antigen CD4. *Cell* 54:66–72.

31. Rosenstein, Y., S. J. Burakoff, and S. H. Herrmann. 1990. HIV-gp120 can block CD4-class II MHC-mediated adhesion. *J. Immunol.* 144:526–531.

32. Sattentau, Q. J., A. G. Dalgleish, R. A. Weiss, and P. C. L. Beverly. 1986. Epitopes of the CD4 antigen and HIV infection. *Science* 234:1120–1123.

33. Schols, D., M. Baba, R. Pauwels, J. Desmyter, and E. de Clercq. 1989. Specific interaction of aurintricarboxylic acid with the human immunodeficiency virus/CD4 cell receptor. *Proc. Natl. Acad. Sci. USA* 86:3322–3326.

34. **Shin, J., C. Doyle, Z. Yang, D. Kappes, and J. L. Strominger.** 1990. Structural features of the cytoplasmic region of CD4 required for internalization. *EMBO J.* **9**:425–434.

35. **Sleckman, B. P., M. Bigby, J. L. Greenstein, S. J. Burakoff, and M.-S. Sy.** 1989. Requirements for modulation of the CD4 molecule in response to phorbol myristate acetate. Role of the cytoplasmic domain. *J. Immunol.* **142**:1457–1462.

36. **Sleckman, B. P., A. Peterson, J. A. Foran, J. C. Gorga, C. J. Kara, J. L. Strominger, S. J. Burakoff, and J. L. Greenstein.** 1988. Functional analysis of a cytoplasmic domain deleted mutant of the CD4 molecule. *J. Immunol.* **140**:49–54.

37. **Sleckman, B. P., A. Peterson, W. K. Jones, J. A. Foran, J. L. Greenstein, B. Seed, and S. J. Burakoff.** 1987. Expression and function of CD4 in a murine hybridoma. *Nature* (London) **328**:351–353.

38. **Smith, D. H., R. A. Byrn, S. A. Marsters, T. Gregory, J. E. Groopman, and D. J. Capon.** 1987. Blocking of HIV-1 infectivity by a soluble, secreted form of the CD4 antigen. *Science* **238**:1704–1707.

39. **Stroynowski, I., M. Soloski, M. G. Low, and L. Hood.** 1987. A single gene encodes soluble and membrane-bound forms of the major histocompatibility Qa-2 antigen: anchoring of the product by a phospholipid tail. *Cell* **50**:759–768.

40. **Traunecker, A., W. Luke, and K. Karjalainen.** 1988. Soluble CD4 molecules neutralize human immunodeficiency virus type 1. *Nature* (London) **331**:84–86.

41. **Traunecker, A., J. Schneider, H. Kiefer, and K. Karjalainen.** 1989. Highly efficient neutralization to HIV with recombinant CD4-immunoglobulin molecules. *Nature* (London) **339**:68–70.

42. **Veillette, A., M. A. Bookman, E. M. Horak, and J. B. Bolen.** 1988. The CD4 and CD8 T cell surface antigens are associated with the internal membrane tyrosine-protein kinase p56[lck]. *Cell* **55**:301–308.

Viruses That Affect the Immune System
Edited by Hung Y. Fan et al.
© 1991 American Society for Microbiology, Washington, DC 20005

Chapter 7

Human Immunodeficiency Virus Entry into Cells

Kathleen A. Page, Nathaniel R. Landau, and Dan R. Littman

The replication cycle of human immunodeficiency virus (HIV) begins with binding of the viral envelope glycoprotein gp160 to its cellular receptor, CD4. Helper T lymphocytes and cells of the monocyte-macrophage lineage express CD4 and are consequently the principal cellular targets of infection. Since CD4 expression is the major host cell determinant for HIV tropism, the interaction of gp160 and CD4 has been a focus for studies on HIV entry and for development of anti-HIV therapies. HIV-infected individuals have also shown evidence of infection in cells thought to be devoid of CD4, including nonhematopoetic cells of the brain and retina (28, 38). A variety of cell lines have been described that do not express detectable levels of CD4, yet are infectable with HIV in vitro. These include various neuronal, hepatic, and glial cells and certain fibroblasts (4, 6, 10, 19, 26, 37). The cellular receptor used by HIV in these cases is unknown. In some cases, HIV infection of monocytes and macrophages may proceed via a receptor other than CD4. In the presence of specific concentrations of antibodies, HIV may enter these cells through the Fc receptor (12, 29). Fc receptor induced by cytomegalovirus infection of fibroblasts is also capable of mediating productive HIV infection in vitro (25). The ability to form "pseudotype" virions with the envelope glycoproteins of other viruses may also expand the host range of HIV. It has

Kathleen A. Page and Nathaniel R. Landau • Department of Microbiology and Immunology, University of California, San Francisco, California 94143-0414. **Dan R. Littman** • Department of Microbiology and Immunology, University of California, San Francisco, California 94143-0414, and Howard Hughes Medical Institute, University of California, San Francisco, California 94143-0724.

recently been shown that HIV is able to form mixed-phenotype virions with other retroviruses and with herpesviruses in vitro (21, 33, 40). The incorporation of non-HIV glycoproteins into its envelope enables HIV to use the cellular receptors of other viruses. While it is clear that HIV predominantly utilizes CD4 as its receptor, it is important to recognize that other mechanisms may also play a role in viral entry.

The events that occur after HIV binding to cells are poorly understood. The viral envelope and cellular membrane must fuse and release the virus core into the cytoplasm. Membrane fusion may occur immediately after virus binding to the cell surface or after internalization of bound virus into an endocytic compartment. Receptor-mediated endocytosis is unlikely to be a requirement for HIV entry since CD4 mutants that are impaired for internalization and mutants that lack cytoplasmic domains serve as efficient receptors for HIV infection (2, 14). Furthermore, HIV infection is not blocked by agents that block the process of vesicle acidification during receptor-mediated endocytosis (23, 35). Cells which express high levels of gp160 will fuse to CD4-bearing cells at neutral pH (20). Since cellular syncytium formation is thought to proceed in a manner analogous to virus-cell fusion, this property of gp160 and CD4 is further evidence that virus fusion occurs immediately upon binding of virus to the plasma membrane.

Membrane fusion is likely to be mediated by a stretch of amino acids located at the N terminus of the transmembrane protein chain of gp160 (Fig. 1). This hydrophobic sequence of amino acids is referred to as the fusion domain and is conserved among retroviruses and some paramyxoviruses (9). Mutations that introduce charged amino acids into the fusion domain and mutations that prevent gp160 from being cleaved into its surface (gp120) and transmembrane (gp41) components abrogate fusion activity (15, 24). The fusion domain may interact directly with the cellular membrane or it may bind to a protein receptor. After fusion of the viral and cellular membranes, the core particle enters the cytoplasm. The postentry processes of reverse transcription and integration can then begin.

Functional and structural aspects of gp160 have been studied in great detail. Some of the salient features of gp160 and CD4 are shown schematically in Fig. 1. The CD4 binding region has been mapped using site-directed mutagenesis to a region near the C terminus of gp120 (18). The affinity of gp160 for CD4 is extremely high, with a binding constant of approximately 10^{-9} M (18, 32). The minimal gp160 binding site of CD4 lies within its N-terminal immunoglobulinlike domain (16). Soluble forms of CD4 that include this domain can effectively compete with cell surface CD4 for HIV binding and thereby block infection (8,13,32). Adjacent to the gp160 binding site on CD4 is a region that appears to be involved in gp160/CD4-mediated fusion. A specific mutation in this region resulted in loss of ability of CD4 to mediate

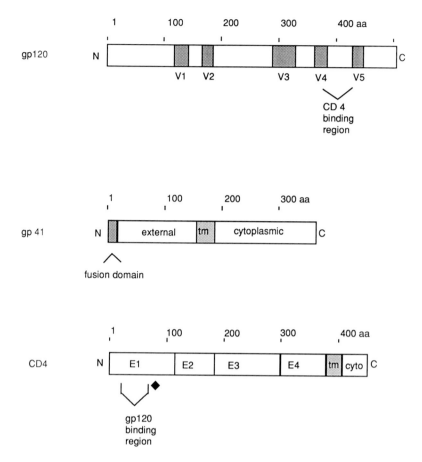

Figure 1. Schematic representation of gp120, gp41, and CD4. External (E), transmembrane (tm), and cytoplasmic (cyto) domains are indicated. The site on CD4 where an introduced mutation resulted in loss of cell-cell fusion activity is indicated with a diamond.

cell-cell fusion without loss of gp160 binding affinity (3). It is possible that after gp160 binding, specific sequences in CD4 interact with gp160 to promote membrane fusion.

N-terminal to the CD4 binding site on gp160 is an exposed domain called variable loop 3 (V3 loop). This domain exhibits considerable sequence variation among HIV isolates: it is the third of five hypervariable regions in gp120 (34). Most neutralizing antibodies bind gp120 at the V3 loop (30). Antibodies specific to this region do not block HIV binding to CD4 but do prevent infection, presumably by interfering with fusion or penetration (31). At the tip of the V3 loop is a conserved sequence of amino acids, GPGRAF, that may serve as a protease cleavage site (36). This conserved sequence is also found in peptides that act as protease inhibitors (11). A recently described trypsinlike protease called tryptase TL is present in T-cell membranes and may be involved in mediating HIV infection and syncytia formation. Antibodies to tryptase TL are reported to block syncytia formation (11). It is tempting to speculate that upon HIV binding to CD4, the V3 loop undergoes a structural change such as cleavage which activates the fusion domain of gp41.

An important aspect of HIV pathogenesis is the marked host range preference of individual viral isolates. Many HIV strains have been classified as either T-cell or macrophage tropic. In two separate studies, when recombinant viruses were prepared by constructing chimeras of viruses with different host ranges, the region responsible for host range specificity was mapped to the envelope gene (5, 39). Since both T-cell- and macrophage-tropic strains use CD4 as their receptor, some property of gp160 other than CD4 binding must be responsible for the host range preference. In another study, a single amino acid change in gp160, introduced by site-directed mutagenesis, was shown to alter HIV cellular tropism without influencing CD4 binding ability (7). One possible interpretation of these results is that, in addition to CD4, gp160 interacts with a cell surface molecule that differs on T cells and macrophages.

Mouse cells that express human CD4 bind HIV but are not permissive for infection and do not fuse with other cells expressing gp160 (1, 22). These cells either lack an essential cellular factor required for infection or have an inhibitor of infection. Cell surface tryptases and fusion domain receptors have been proposed as potential factors that may be required to allow infection. The block to HIV infection in mouse cells provides a starting point for identification of cellular components that are involved in infection.

USE OF AN HIV VECTOR TO STUDY VIRUS ENTRY

To facilitate studies on the mechanism of HIV entry into cells, we constructed an HIV vector that would permit selection of HIV-infectable cells (27). A drug resistance gene was inserted into a gp160-negative HIV proviral genome such that it could be packaged into HIV virions. The HIV genome was rendered replication defective by deletion of sequences encoding gp160 and insertion of a *gpt* gene with a simian virus 40 promoter at the deletion site. This proviral construct is called HIV-gpt. Cotransfection of HIV-gpt with a gp160 expression construct into Cos-7 cells results in packaging of the defective HIV-gpt genome into infectious virions. The virus preparations derived from the Cos-7 cells can then be titered on a variety of host cells. The drug resistance gene is transmitted to and expressed in infectable cells, enabling their selection in mycophenolic acid. This system, shown schematically in Fig. 2, provides a quantitative measure of HIV infection, since each successful infection event leads to the growth of a drug-resistant colony. When transfected in the absence of an envelope expression construct, envelope-deficient virions are generated. These virions contain HIV-gpt genomic RNA but are not infectious (27). The ability of retroviruses to form pseudotypes with the envelope glycoproteins of other viruses can be readily exploited by HIV-gpt. When a heterologous *env* gene such as one from an amphotropic murine leukemia virus (A-MLV) is cotransfected with HIV-gpt, viruses are produced that have an HIV core and A-MLV envelope glycoprotein. This pseudotype virus has the host range of A-MLV (27). HIV-gpt has also been pseudotyped with the envelope glycoprotein of human T-cell leukemia virus type I (17). The ability of HIV to utilize this envelope glycoprotein may have important implications for the spread of HIV in individuals infected with both viruses. The use of HIV-gpt to prepare pseudotype viruses has certain advantages over the more traditional method of mixed infections. The envelope protein is the only heterologous component in the pseudotype, and the homologous envelope is completely absent. Thus, pseudotypes prepared by this method are biochemically and biologically well defined. Studies on the role of the *env* gene product are facilitated by HIV-gpt since *env* is supplied on a separate vector. The effect of gp160 on HIV host range preference can be studied by complementing HIV-gpt with *env* genes derived from various HIV isolates and examining changes in host range.

HIV-gpt produced in the presence of gp160 derived from HIV strain HXB2 was designated HIV(HXB2). This virus was tropic for CD4$^+$ human cells and, like wild-type HIV, could not infect 3T3-T4 cells (Table 1). Infection of HeLa T4 cells with this virus was blocked by soluble CD4 (data not shown). Expression of an A-MLV envelope gene in cells transfected with HIV-gpt led to the production of virus [designated HIV(A-MLV)] capable of infecting both human and murine cells (Table 1). These results conform to our previously reported results (27) and demonstrate that the host range properties of viruses prepared with HIV-gpt accurately reflect the host range of the envelope gene employed. Since HIV(A-MLV) readily infected 3T3-T4 cells, these cells must be permissive for HIV DNA synthesis and integration. Thus, the block to HIV infection of murine cells must occur during a gp160-mediated event during viral entry.

To determine whether gp160 genes from different viral isolates varied in their host range properties, *env* genes from two HIV isolates, HXB2 and SF33, were cloned into an expression construct. These *env* genes were expressed in Cos-7 cells with HIV-gpt, and the virus produced was titered on HeLa T4, 3T3-T4, and SK-N-MC cells. SK-N-MC is a neuroblastoma cell line previously reported to be susceptible to HIV infection (19). Since indi-

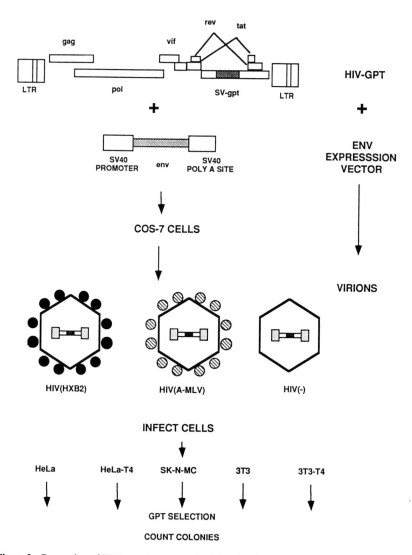

Figure 2. Generation of HIV-gpt viruses. A 1.2-kb deletion in the *env* gene of HIV strain HXB2 was made, leaving the *rev* response element and *rev* and *tat* exons intact. Simian virus 40 (SV40) and *gpt* sequences were inserted at the deletion site. The *env* expression construct contains the complete coding region for gp160 but does not include translation initiation codons for *rev, tat,* or *vpu.* gp160 is the only HIV gene product supplied by the *env* expression vector; all other viral gene products are derived from HIV-gpt. HIV-gpt and *env* expression constructs were transfected into Cos-7 cells as previously described (27). At 48 to 78 h after transfection the culture supernatant was harvested and filtered through a 0.4 μm-pore-size filter. The virus-containing culture supernatants were stored at −70°C. To titrate infectivity, the virus preparations were diluted serially and added to target host cells. The next day, the virus inoculum was removed and infected cells were selected in media containing mycophenolic acid as previously described (27). Mycophenolic acid-resistant colonies were stained and counted after 10 to 16 days. LTR, long terminal repeat.

Table 1. Infection of HeLa T4 and
3T3-T4 Cells with HIV-gpt

Virus prepn	Infectious units/ml	
	HeLa T4	3T3-T4
HIV(HXB2)	3,500	0
HIV(A-MLV)	163	2,800

vidual HIV isolates vary considerably in their ability to infect this cell line, it can be used as a target cell line for identifying and characterizing potential host range differences conferred by different HIV *env* genes. As a control, HIV-gpt virions without any *env* protein were also prepared [designated HIV(−)]. The levels of virus production in the transfected Cos-7 cells were equivalent for all three virus preparations: 60 ng of p24 per ml for HIV(HXB2), 55 ng of p24 per ml for HIV(SF33), and 57 ng of p24 per ml for HIV(−). The level of envelope gene expression in the HIV(HXB2) and HIV(SF33) preparations was also very similar when analyzed by immunoblotting (data not shown). As shown in Table 2, HIV(HXB2) and HIV(SF33) varied in their host range properties. Whereas the two viruses infected SK-N-MC cells equivalently, HIV(SF33) had a greatly reduced titer on HeLa T4 cells. Neither virus could infect 3T3-T4 cells. Predictably, envelope-deficient HIV(−) was unable to mediate infection of any cell type. This result underscores the absolute requirement for an envelope glycoprotein in order to initiate retroviral infection. The reason for the differing infectivity of HIV(HXB2) and HIV(SF33) for HeLa T4 cells is unknown but must be related to some property of gp160.

HIV-gpt infection of SK-N-MC cells may proceed via CD4-dependent and CD4-independent mechanisms. These cells did not express detectable levels of CD4 when tested by surface staining with the anti-CD4 antibody Leu3A, but this technique may not be adequately sensitive to detect rare CD4-positive cells in the population. When HIV(HXB2) and HIV(SF33) were treated with soluble CD4 before infection of SK-N-MC cells, infection was

Table 2. Influence of gp160 Strain Variations
on Infectivity

Virus prepn	Infectious units/ml		
	HeLa T4	SK-N-MC	3T3-T4
HIV(HXB2)	1,600	32	0
HIV(SF33)	290	48	0
HIV(−)	0	0	0

inhibited 90 to 100%. Treatment of SK-N-MC cells with the anti-CD4 antibody OKT4A resulted in 90 to 95% inhibition of infection. Infection of SK-N-MC cells with HIV(A-MLV) was unaffected by either soluble CD4 or OKT4A antibody treatment. These results suggest that HIV(HXB2) and HIV(SF33) infection of SK-N-MC cells is principally mediated by CD4 but some CD4-independent viral entry may also occur. SK-N-MC infection with HIV strain HTLV-IIIRF was previously reported to be insensitive to soluble CD4 and anti-CD4 antibody treatment (19). The reason for the discrepancy between our results and those previously reported is unclear but may be related to the strain of HIV used and the specific cell and virus culture conditions employed.

In summary, HIV-gpt is a self-packaging retroviral vector that can be used to quantitate viral infection. Studies on the role of *env* are facilitated by HIV-gpt since *env* is supplied on a separate vector. HIV-gpt can be complemented with retroviral *env* genes other than gp160 such that pseudotype viruses are readily generated. We have used HIV-gpt to determine that mouse 3T3-T4 cells are blocked for HIV entry at the level of a gp160-mediated event but do permit postentry events of DNA synthesis and integration. We have also shown that strain variations in gp160 may influence the host range preferences of HIV. HIV-gpt can be used to select for cells permissive for the early events of HIV infection since stable *gpt* expression is used as the marker for infection. Thus, HIV-gpt allows analysis of a unique window of virus infection events up to and including integration. The use of HIV-gpt as a tool for understanding HIV-host cell interactions may lead to the identification of cellular factors involved in HIV infection.

Acknowledgments. We thank Cecelia Cheng-Mayer and Jay A. Levy for providing an SF33 construct and Gail Mosley for help in preparing the manuscript. This work was supported by National Institutes of Health grants AI 0890 to K.A.P. and AI 24286 to D.R.L.

REFERENCES

1. **Ashorn, P. A., E. A. Berger, and B. Moss.** 1990. Human immunodeficiency virus envelope glycoprotein/CD4 mediated fusion of nonprimate cells with human cells. *J. Virol.* **64:**2149–2156.
2. **Bedinger, P., A. Moriarty, R. C. von Borstel, N. J. Donovan, K. S. Steimer, and D. R. Littman.** 1988. Internalization of the human immunodeficiency virus does not require the cytoplasmic domain of CD4. *Nature* (London) **334:**162–165.
3. **Camerini, D., and B. Seed.** 1990. A CD4 domain important for HIV-mediated syncytium formation lies outside the virus binding site. *Cell* **60:**747–754.
4. **Cao, Y., A. E. Friedman-Kein, Y. Huang, X. L. Li, M. Mirabile, T. Moudgil, D. Zucker-Franklin, and D. Ho.** 1990. CD4-independent, productive human immunodeficiency virus type 1 infection of hepatoma cell lines in vitro. *J. Virol.* **64:**2553–2559.
5. **Cheng-Mayer, C., M. Quiroga, J. W. Tung, D. Dina, and J. A. Levy.** 1990. Viral determinants

of human immunodeficiency virus type 1 T-cell or macrophage tropism, cytopathogenicity, and CD4 antigen modulation. *J. Virol.* **64**:4390–4398.

6. **Cheng-Mayer, C., J. T. Rutka, M. L. Rosenblum, T. McHugh, and D. P. Stites, and J. A. Levy.** 1987. Human immunodeficiency virus can productively infect cultured human glial cells. *Proc. Natl. Acad. Sci. USA* **84**:3526–3530.

7. **Cordonnier, A., L. Montagnier, and M. Emerman.** 1989. Single amino-acid changes in HIV envelope affect viral tropism and CD4 binding. *Nature* (London) **340**:571–574.

8. **Fisher, R. A., J. M. Bertonis, W. Meier, V. A. Johnson, D. S. Costopoulos, T. Liu, R. Tizard, B. D. Walker, M. S. Hirsch, R. T. Schooley, and R. A. Flavell.** 1988. HIV infection is blocked in vitro by recombinant soluble CD4. *Nature* (London) **331**:76–78.

9. **Gallaher, W. R.** 1987. Detection of a fusion peptide sequence in the transmembrane protein of human immunodeficiency virus. *Cell* **50**:327–328.

10. **Harous, J. M., C. Kunsch, H. T. Hartle, M. A. Laughlin, J. A. Hoxie, B. Wigdahl, and F. Gonzalez-Scarano.** 1989. CD4 independent infection of human neural cells by human immunodeficiency virus type 1. *J. Virol.* **63**:2527–2533.

11. **Hattori, T., A. Koito, K. Takatsuki, H. Kido, and N. Katunuma.** 1989. Involvement of tryptase-related cellular protease(s) in human immunodeficiency type 1 infection. *FEBS Lett.* **248**:48–52.

12. **Homsy, J., M. Meyer, M. Tateno, S. Clarkson, and J. A. Levy.** 1989. The Fc and CD4 receptor mediates antibody enhancement of HIV infection of human cells. *Science* **244**:1357–1360.

13. **Hussey, R. E., N. E. Richardson, M. Kowalski, N. R. Brown, H. S. Chang, R. F. Siliciano, T. Dorfman, B. Walker, J. Sodroski, and E. L. Reinherz.** 1988. A soluble CD4 protein selectively inhibits HIV replication and syncytium formation. *Nature* (London) **331**:78–82.

14. **Jasin, M., K. A. Page, and D. R. Littman.** 1991. GPI-anchored CD4/Thy-1 chimeric molecules serve as HIV receptors in human, but not mouse, cells and are modulated by gangliosides. *J. Virol.* **65**:440–444.

15. **Kowalski, M., J. Potz, L. Basiripour, T. Dorfman, W. C. Goh, E. Terwilliger, A. Dayton, C. Rosen, W. Haseltine, and J. Sodroski.** 1987. Functional regions of the envelope glycoprotein of human immunodeficiency virus type 1. *Science* **237**:1351–1355.

16. **Landau, N. R., M. Warton, and D. R. Littman.** 1988. The envelope glycoprotein of the human immunodeficiency virus binds to the immunoglobulin-like domain of CD4. *Nature* (London) **334**:159–162.

17. **Landau, N. R., K. A. Page, and D. R. Littman.** 1991. Pseudotyping with human T-cell leukemia virus type I broadens the human immunodeficiency virus host range. *J. Virol.* **65**:162–169.

18. **Lasky, L. A., G. Nakamura, D. H. Smith, C. Fennie, C. Shimasaki, E. Patzer, P. Berman, T. Gregory, and D. J. Capon.** 1987. Delineation of a region of the HIV-1 gp120 glycoprotein critical for interaction with the CD4 receptor. *Cell* **50**:975–985.

19. **Li, X. L., T. Moudgil, H. V. Vinters, and D. Ho.** 1990. CD4-independent, productive infection of a neuronal cell line by human immunodeficiency virus type 1. *J. Virol.* **64**:1383–1387.

20. **Lifson, J. D., M. B. Feinberg, G. R. Reyes, L. Rabin, B. Banapour, S. Chakrabarti, B. Moss, F. Wong-Staal, K. S. Steimer, and E. G. Engleman.** 1986. Induction of CD4-dependent cell fusion by the HTLV-III/LAV envelope glycoprotein. *Nature* (London) **323**:725–728.

21. **Lusso, P., F. M. Veronese, B. Ensoli, G. Franchini, C. Jemma, S. E. DeRocco, V. S. Kalyanaraman, and R. C. Gallo.** 1990. Expanded HIV-1 cellular tropism by phenotypic mixing with murine endogenous retroviruses. *Science* **247**:848–852.

22. **Maddon, P. J., A. G. Dalgleish, J. S. McDougal, P. R. Clapham, R. A. Weiss, and R. Axel.** 1986. The T4 gene encodes the AIDS virus receptor and is expressed in the immune system and the brain. *Cell* **47**:333–348.

23. **McClure, M. O., M. Marsh, and R. A. Weiss.** 1988. Human immunodeficiency virus infection of CD4-bearing cells occurs by a pH-independent mechanism. *EMBO J.* 7:513–518.

24. **McCune, J. M., L. B. Rabin, M. B. Feinberg, M. Lieberman, J. C. Kosek, G. R. Reyes, and I. L. Weissman.** 1988. Endoproteolytic cleavage of gp160 is required for activation of human immunodeficiency virus. *Cell* 53:55–67.

25. **McKeating, J. A., P. D. Griffiths, and R. A. Weiss.** 1990. HIV susceptibility conferred to human fibroblasts by CMV-induced Fc receptor. *Nature* (London) 343:659–661.

26. **Mellert, W., A. Kleinschmidt, J. Schmidt, H. Festl, S. Emler, W. K. Roth, and V. Erfle.** 1990. Infection of human fibroblasts and osteoblast-like cells with HIV-1. *AIDS* 4:527–536.

27. **Page, K. A., N. R. Landau, and D. R. Littman.** 1990. Construction and use of a human immunodeficiency virus vector for analysis of virus infectivity. *J. Virol.* 64:5270–5276.

28. **Pomerantz, R. J., D. R. Kuritzkes, S. M. de la Monte, T. R. Rota, A. S. Baker, D. Albert, D. H. Bor, E. L. Feldman, R. T. Schooley, and M. S. Hirsch.** 1987. Infection of the retina by human immunodeficiency virus type 1. *N. Engl. J. Med.* 317:1643–1647.

29. **Robinson, W. E., D. C. Montefiori, and W. M. Mitchell.** 1988. Antibody-dependent enhancement of human immunodeficiency virus type 1 infection. *Lancet* i:790–794.

30. **Rusche, J. R., K. Javaherian, C. McDanal, J. Petro, D. L. Lynn, P. J. Fischinger, D. P. Bolognesi, S. D. Putney, and T. J. Matthews.** 1988. Antibodies that inhibit fusion of HIV-1 infected cells bind a 24 amino acid sequence of the viral envelope, gp120. *Proc. Natl. Acad. Sci. USA* 85:3198–3202.

31. **Skinner, M. A., A. J. Langlois, C. B. McDanal, J. S. McDougal, D. P. Bolognesi, and T. J. Matthews.** 1988. Neutralizing antibodies to an immunodominant envelope sequence do not prevent gp120 binding to CD4. *J. Virol.* 62:4195–4200.

32. **Smith, D. H., R. A. Byrn, S. A. Marsters, T. Gregory, J. E. Groopman, and D. J. Capon.** 1987. Blocking of HIV-1 infectivity by a soluble, secreted form of the CD4 antigen. *Science* 238:1704–1707.

33. **Spector, D. H., E. Wade, D. A. Wright, V. Koval, C. Clark, D. Jaquish, and S. A. Spector.** 1990. Human immunodeficiency virus pseudotypes with expanded cellular and species tropism. *J. Virol.* 64:2298–2308.

34. **Starcich, B. R., B. H. Hahn, G. M. Shaw, P. D. McNeely, S. Modrow, H. Wolf, E. S. Parks, W. P. Parks, S. F. Josephs, R. C. Gallo, and F. Wong-Staal.** 1986. Identification and characterization of conserved and variable regions in the envelope gene of HTLV-III/LAV, the retrovirus of AIDS. *Cell* 45:637–648.

35. **Stein, B. S., S. D. Gowda, J. D. Lifson, R. C. Penhallow, K. G. Bensch, and E. G. Engleman.** 1987. pH-independent HIV entry into CD4-positive T cells via virus envelope fusion to the plasma membrane. *Cell* 49:659–668.

36. **Stephens, P. E., G. Clements, G. T. Yarranton, and J. Moore.** 1990. A chink in HIV's armour? *Nature* (London) 343:219. (Letter.)

37. **Tateno, M., F. Gonzalez-Scarano, and J. A. Levy.** 1989. Human immunodeficiency virus can infect CD4-negative human fibroblastoid cells. *Proc. Natl. Acad. Sci. USA* 86:4287–4290.

38. **Wiley, C. A., R. A. Schrier, J. A. Nelson, P. W. Lampert, and M. B. A. Oldstone.** 1986. Cellular localization of human immunodeficiency virus infection within the brains of acquired immunodeficiency syndrome patients. *Proc. Natl. Acad. Sci. USA* 83:7089–7093.

39. **York-Higgins, D., G. Cheng-Meyer, D. Bauer, J. A. Levy, and D. Dina.** 1990. Human immunodeficiency virus type 1 cellular host range, replication, and cytopathicity are linked to the envelope region of the viral genome. *J. Virol.* 64:4016–4020.

40. **Zhu, Z., S. S. L. Chen, and A. S. Huang.** 1990. Phenotypic mixing between human immunodeficiency virus and vesicular stomatitis virus or herpes simplex virus. *AIDS* 3:215–219.

Viruses That Affect the Immune System
Edited by Hung Y. Fan et al.
© 1991 American Society for Microbiology, Washington, DC 20005

Chapter 8

The Murine Acquired Immunodeficiency Syndrome (MAIDS) Induced by the Duplan Strain Retrovirus

Paul Jolicoeur, Zaher Hanna, Doug Aziz, Carole Simard, and Ming Huang

The acquired immunodeficiency syndrome (AIDS) is one of the few human diseases in which a retrovirus, the human immunodeficiency virus (HIV), appears to play a major role (10, 56). Despite a large body of information which has been accumulated on the virus growth cycle and on the viral gene products, the mechanism by which HIV contributes to the development of a progressive immunodeficiency after a long latent period remains obscure. It seems that good animal models will be required to understand the pathogenesis of AIDS and to develop efficient therapies and vaccines. The immunodeficiencies induced by the lentiviruses such as the simian (8) and feline (46) immunodeficiency viruses appear to mimic many features of the human disease and therefore represent very good animal models for AIDS. Other retroviruses of the non-lentivirus class, such as the simian type D retrovirus (35) or the defective feline (40) and murine (2, 6) leukemia viruses, have been shown to induce severe immunodeficiency in their respective hosts and may contribute significantly to our understanding of the pathogenesis of retrovirus-induced immunodeficiencies. The different features of these animal

Paul Jolicoeur • Laboratory of Molecular Biology, Clinical Research Institute of Montreal, 110 Pine Avenue West, Montreal, Quebec, Canada H2W 1R7, and Département de Microbiologie et d'Immunologie, Faculté de Médecine, Université de Montréal, Montreal, Quebec, Canada H3C 3J7.   **Zaher Hanna, Doug Aziz, Carole Simard, and Ming Huang** • Laboratory of Molecular Biology, Clinical Research Institute of Montreal, 110 Pine Avenue West, Montreal, Quebec, Canada H2W 1R7.

models have been reviewed (8, 12). Recently, a better understanding of the murine AIDS (MAIDS) has emerged, and we would like to review our present knowledge on this disease in more detail.

DISCOVERY OF THE MOUSE RETROVIRUS INDUCING MAIDS

The discovery of the MAIDS virus was fortuitous and was not recognized as such for many years. While trying to reproduce data from Lieberman and Kaplan (32) on the induction of thymomas by cell-free extracts from X-ray-induced thymic lymphomas of C57BL/6 mice, Latarget and Duplan could indeed reproduce these data, obtaining extracts inducing thymic lymphomas (27). However, they noticed that in nearly 50% of the C57BL/6 mice, the thymus was not involved and mice developed enlarged spleens and lymph nodes (36). From 1961 to 1964, they passaged the virus by cell-free extracts of spleens and lymph nodes. Between 1965 and 1969, they finally obtained a crude virus preparation which reproducibly induced enlarged lymph nodes and spleen in 100% of C57BL/6 mice: with this extract 25 to 50% of inoculated mice also exhibited thymomas (36). In 1971, the cell-free extract became highly virulent in a rather short time (36). The investigators realized that the disease was clearly distinct from thymic lymphoma by the observation that it could be induced in young adult mice, that thymectomy did not prevent its development, and that several attempts to transplant cells from these enlarged nodes or spleens failed (36). Duplan's group thought that these mice had lymphoreticular tumors, resembling type B reticulum cell sarcoma (28, 36), but they could not identify the "cell population whose proliferation (was) responsible for tumor formation" (28). They showed that these enlarged organs contained a heterogenous population of T cells, B cells, and non-T/non-B cells and that B cells and non-T/non-B cells proliferated constantly (28). They also found an almost total loss of T-cell response to the mitogens concanavalin A and phytohemagglutinin, a decrease of B-cell response to lipopolysaccharide, and hypergammaglobulinemia (immunoglobulin M and immunoglobulin G2a) (28). These are major features of an immunodeficiency. Duplan's group reported their data in the context of cancer, before AIDS was largely known. They did not recognize that the immunodeficiency they had observed had characteristics similar to those of other immunodeficiency syndromes.

Five years later, at a time when AIDS was now a widely discussed disease, Mosier et al. (38) repeated several of the experiments reported by Legrand et al. (28) and confirmed them, but surprisingly did not seem aware of Legrand's work and did not quote it. These investigators also extended this work, most importantly by realizing that the immunodeficiency seen in these mice shared common features with human AIDS (37, 38). For this reason, the disease induced by this virus was designated murine AIDS (MAIDS) (37, 38).

IDENTIFICATION OF THE ETIOLOGIC AGENT OF MAIDS

Attempts To Identify the Virus with Classical Virological Techniques

Several attempts by Duplan's group to isolate the etiologic agent of the disease failed (1, 15, 29). Most of the techniques then used involved virus cloning at limiting dilution and screening for replication-competent retroviruses. Several classes of murine leukemia viruses (MuLVs) were isolated, some of them inducing typical B-cell lymphomas after a long latent period (1, 15, 29, 33). Our present knowledge of the defective nature of the MAIDS virus (2, 6) helps us to understand retrospectively why these experiments failed: they were designed to screen against the pathogenic virus.

Other attempts involved the establishment of cell lines in vitro which could produce the pathogenic agent. Such cell lines were first obtained (33), but eventually stopped producing the pathogenic agent (16). However, Haas and Meshorer were able to establish several stromal hematopoietic cell lines producing a highly pathogenic virus (17). The disease induced by these viruses produced in vitro (designated most frequently as LP-BM5) was identical to the disease induced by the crude virus preparation (17). Further attempts by classical virologic methods to identify and isolate the etiologic agent of the disease produced by these cells were unsuccessful, although several different MuLVs of various classes were identified (18). Altogether, a large number of distinct retroviruses (MuLV) were isolated from cell lines or crude extracts of diseased tissues (3 ecotropic B-tropic MuLVs [1, 15, 18, 29], 1 ecotropic N-tropic MuLV [33], 1 xenotropic MuLV [1, 18, 33], and 16 dualtropic MuLVs [18]), but none of them induced MAIDS. This emphasizes the notion that in retrovirus-induced disease, diseased tissues often contain nonpathogenic retroviruses. The situation with AIDS may well be similar, and many of the HIV isolates obtained in tissue culture may not be pathogenic.

Evidence that the MAIDS Virus Is Defective

The difficulty in isolating the pathogenic agent of MAIDS suggested that it was unique and might be defective: some authors had already suggested such a hypothesis (18). We therefore used a different strategy for our study, especially avoiding biological cloning of the virus. We obtained a crude virus preparation (an extract from a lymph node of a MAIDS mouse) from Duplan's laboratory in October 1985. The extract induced typical disease in our

C57BL/6 mice. This crude extract was used to infect SC-1 mouse fibroblasts in the presence of Polybrene. After a few days, this culture produced high levels of MuLV, as determined by a reverse transcriptase assay, and the supernatant was used to infect young adult C57BL/6 mice. This supernatant was found to contain viruses able to induce MAIDS after a relatively short latent period. We used this supernatant to reinfect SC-1 cells and to analyze the unintegrated viral DNA in a Hirt supernatant. To our surprise and delight we found a viral DNA species which was not usually detected with other preparations of leukemogenic MuLVs. It was a linear viral DNA almost half the size of a typical viral genome: a defective viral DNA (2). Surprisingly, the levels of this viral DNA were almost identical to those of the helper MuLVs. Using a similar approach, Chattopadhyay et al. (6) also detected a defective viral DNA genome in their preparation of the pathogenic LP-BM5 Duplan virus. The presence of this defective viral genome in virus preparations was found by this group (6) and us (2) to correlate with the ability of the extracts to induce MAIDS.

Since this retrovirus species was so unusual, we hypothesized that it could be involved in the disease and we decided to clone its genome molecularly. The unintegrated supercoiled defective viral DNA (4.8 kbp), enriched by Hirt extraction and on a CsCl-propidium iodide gradient, was cleaved at its unique *Hind*III site and cloned in Charon 4A (2).

This defective viral DNA was transfected into SIM.R mouse fibroblasts, and the virus was rescued with a nonpathogenic helper MuLV. This molecularly cloned virus induced MAIDS (2) (Fig. 1 and 2), indicating that it was the pathogenic agent of the disease, thus fulfilling Koch's postulates. (These results were presented at the Cold Spring Harbor RNA Tumor Viruses Meeting, May 1988.)

Sequencing of this viral genome revealed that it was related to MuLV, had sustained deletions in *pol* and *env*, had a long open reading frame corresponding to a putative *gag* precursor protein, and that the p12 portion of *gag* was the most divergent region (40 to 50%) of the genome as compared to that of other MuLVs (2) (Fig. 3).

The *gag* Precursor Pr60*gag* Is a Gene Product of the MAIDS Virus

Since our sequencing data indicated that a long open reading frame encoding a putative *gag* precursor protein was present in the defective MAIDS virus (2) (Fig. 4), we searched for the presence of this protein in infected fibroblasts. As previously reported (6), we found a novel Pr60*gag* precursor in infected fibroblasts which could be precipitated with antisera against *gag* p15 (MA), p30 (CA), and p10 (NC) proteins, but not with an antiserum against *gag* p12 (21) (Fig. 5). This result correlated with our sequencing data showing

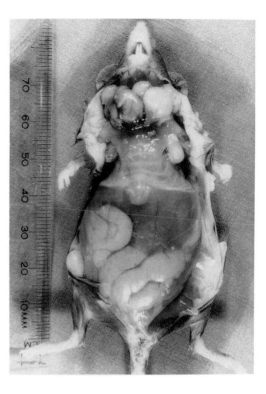

Figure 1. A C57BL/6 mouse showing signs of MAIDS. This mouse was inoculated with the molecularly cloned defective virus (Du5H) pseudotyped with the nonpathogenic G6T2 helper MuLV. Note the severe lymphodenopathy.

that the p12 protein region was the most divergent and was shorter than that of nondefective MuLV by eight amino acid residues.

Further analysis of defective MAIDS virus Pr60^{gag} revealed that it was myristylated, like most *gag* polyproteins of mammalian retroviruses (53), and phosphorylated (21). Subcellular fractionation of this precursor protein showed that it was located in the cell membrane fraction (21), like the myristilated Pr65^{gag} of helper MuLV. Using an antibody specific to the defective MAIDS virus p12 domain which did not cross-react with the p12 from various helper MuLVs, we found that the Pr60^{gag} protein was unable by itself to form virus particles, but could form phenotypically mixed particles with helper MuLV. In these particles, Pr60^{gag} was only partially cleaved, presumably in *trans* by the helper protease, and interfered with proper cleavage of the helper *gag* precursor, generating a *gag*-related intermediate of 40 kDa usually not seen in wild-type virions (21) (Fig. 5).

Together, these data (21) convincingly showed that the Pr60 *gag*/fusion protein was a major gene product of the defective MAIDS MuLV. It may also represent its only gene product since the rest of the genome represents only remnants of deleted *pol* and *env* read in alternate reading frames. Since

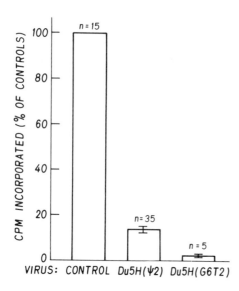

Figure 2. Loss of T-cell function in MAIDS. The response of spleen T cells to the mitogen concanavalin A was measured as described (2, 22) in control uninoculated and in diseased mice inoculated with the molecularly clone-defective (Du5H) MAIDS virus in the absence (ψ2) or in the presence (G6T2) of a replication-competent helper MuLV. Histogram represents maximal activity in counts per minute. Bars represent the standard deviation of the mean for each group.

this defective virus is pathogenic, its major gene product, Pr60^{gag}, is likely to be involved in the pathogenesis of the disease. In fact, recent results on deletion and point mutations of the defective viral genome indicate that indeed Pr60^{gag} is essential for the pathogenicity of the virus (20). The role of Pr60^{gag} may be that of an oncoprotein (see below).

Role of Helper MuLV in Induction of MAIDS by Duplan MuLV

A defective retrovirus needs a helper virus to replicate. In vitro, the helper virus does not seem to be very important, at least for the defective

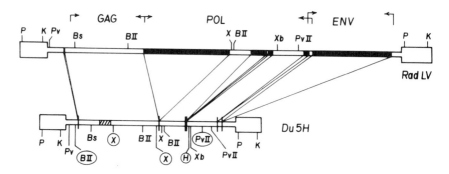

Figure 3. Physical map of the defective MAIDS virus. For comparison, the map of a nondefective MuLV, a radiation leukemia virus (RadLV), is shown. Du5H designates the defective virus clone. Open box, homologous regions; solid box, deleted sequences; hatched box, unique sequences. BII, *Bgl*II; Bs, *Bst*EII; K, *Kpn*I; P, *Pst*I; Pv, *Pvu*I; PvII, *Pvu*II; X, *Xho*; Xb, *Xba*I.

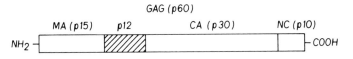

Figure 4. Schematic representation of the Pr60 *gag*/fusion protein. This uncleaved precursor protein encodes domains highly homologous to known *gag* p15 (MA), p30 (CA), and p10 (NC) proteins (open boxes) and to a less conserved domain related to *gag* p12 protein (hatched box).

viruses harboring an oncogene, since they have been shown to transform fibroblasts in the absence of helper retroviruses, that is to say, in the absence of virus replication (3). However, in vivo, the helper viruses have been found to play a critical role when pseudotyped with some defective retroviruses, while they were found to be dispensable with others. An appropriate helper virus was found to be critical for the induction of pre-B-cell lymphoma by the defective Abelson virus, at least in some mice (49–51). However, when T-cell leukemias (thymomas) are induced with the same Abelson virus by intrathymic inoculation, the helper virus seems to play a minor role and does not even appear to be required (47). Similarly, the helper virus is not required to induce erythroleukemia with the defective spleen focus-forming virus (55).

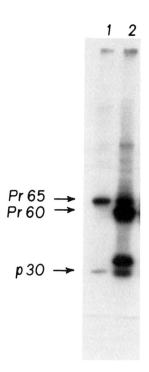

Figure 5. Detection of Pr60*gag* in cells infected with the defective MAIDS virus. Normal ψ2 fibroblasts (lane 1) or ψ2 fibroblasts infected with the defective MAIDS virus (lane 2) were labeled with [³⁵S]methionine. The labeled proteins were precipitated with anti-*gag* p30 antiserum as described (21).

In light of these results, we decided to study the role of virus replication in the induction of MAIDS by the defective virus. To achieve this goal, we constructed helper-free stocks of MAIDS defective virus by transfecting the defective viral genome in the Ψ2 packaging cell line (34). This cell line contains all the helper Moloney MuLV proteins to pseudotype the defective viral genome. However, the Moloney viral genome encoding these proteins is itself deleted of its packaging sequences and therefore cannot be packaged. Ψ2 cells harboring a defective virus, such as the MAIDS virus, produce infectious defective viruses. However, cells infected by these viruses remain nonproducers (34).

We selected clones producing high titers of this helper-free defective MAIDS virus and inoculated young adult C57BL/6 mice with these viruses (22). Interestingly, these stocks were as efficient in inducing typical MAIDS as the replication competent stocks (22) (Table 1, Fig. 2). Every effort to detect replicating MuLVs of different classes in these mice, by cocultivation of diseased tissues on susceptible cells or by RNA detection techniques, was unsuccessful, except for xenotropic MuLVs, which were also present in normal mice. This class of MuLV is thought not to replicate in the mouse (31).

Therefore, it appears that this defective MAIDS virus can induce disease in the absence of virus replication. This is a remarkable result considering the small virus inoculum, the route of inoculation, and the target cells involved. Indeed, we estimated that $\sim 10^4$ infectious virus particles (as titered on mouse fibroblasts by in situ hybridization) were inoculated intraperitoneally. Presumably, these viruses are absorbed and transported systemically and finally reach infectable cells. Since not all infected cells represent the target cells of this virus and since virus inactivation is likely to occur in vivo, it seems reasonable to estimate that at best 10^3 target cells of the virus are initially infected. Considering the number of potentially infectable cells in vivo, it would be surprising to get even that many target cells infected. Therefore, if a relatively large number of target cells get infected by a single injection, it suggests that these cells must be relatively abundant or are concen-

Table 1. Effects of Helper-Free Stocks of Defective MAIDS Virus in C57BL/6 Mice

Virus inoculated[a]	No. of mice diseased/ no. studied	Spleen wt (mg)	Lymph node wt (mg)	Latency (weeks)
No virus	0/25	75–134	10–25	
Du5H (Ψ2)	68/72	87–1,070	114–4,100	12–15
Du5H (G6T2)	43/43	176–1,010	95–918	12–15

[a]Viruses were inoculated intraperitoneally into young (30- to 40-day-old) adult C57BL/6 mice as described (2, 22). The defective MAIDS virus (Du5H) was pseudotyped with a helper-free (Ψ2) or a replication-competent helper (G6T2) MuLV.

trated in a single site such as the peritoneum, or both. Interestingly, these initially few infected target cells are sufficient to eventually kill the animal. How do they achieve this result?

Clonal Expansion of Infected Cells in MAIDS

The surprising result that MAIDS develops with helper-free stocks of the defective Duplan virus, in the absence of virus replication, has given us a tool to study the pathogenesis of the disease, first by studying the fate of the infected target cells.

Using in situ hybridization with a probe able to detect the defective viral RNA, we estimated the number of infected cells in enlarged lymph nodes or spleen of MAIDS mice. Positive cells with numerous grains (Fig. 6) seemed to constitute a variable proportion of cells in these organs (22). A caveat to this experiment is the fact that the probe used also hybridized to endogenous sequences and might have hybridized to induced endogenous viral RNA. However, this result was confirmed by analysis of the proviral defective DNA in these organs. The 4.2-kbp internal *Pst*I proviral DNA fragment was found in all infiltrated organs, but again at a variable level, suggesting that the number of infected cells present in the enlarged nodes varied from one to the next in the same mouse (22).

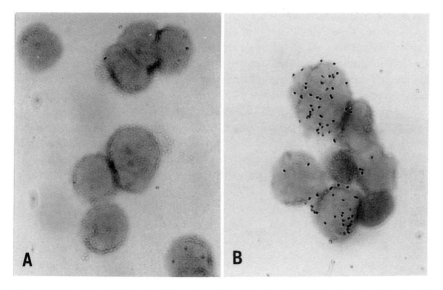

Figure 6. Detection of infected cells in enlarged lymph nodes of MAIDS mice. Lymph node cells were dispersed, collected by centrifugation, and processed for in situ hybridization with ^{32}P-D30 DNA probe as described (22, 44). (A) Spleen cells from a normal uninoculated mouse. (B) Lymph node cells from a diseased mouse. Note two representative cells with numerous grains.

Assuming that 0.1 g of lymph node represents 7×10^7 to 8×10^7 cells and that a diseased mouse 4 months after inoculation harbors about 2 g of enlarged lymph nodes and spleen, we calculated that each diseased mouse harbors about 1.5×10^9 cells within its enlarged diseased tissues (22). If 5% of these cells are infected with the defective virus, as detected in some nodes, and assuming one proviral copy per cell, then our results indicate that a minimum of 7×10^7 cells harbor the defective MAIDS viral genome. Since only 3×10^4 infectious units of the defective virus were initially inoculated into each mouse in two injections, the maximum number of target cells initially infected was 3×10^4. Therefore, our data indicate that at least 1,000-fold more infected cells are present in diseased mice than in newly inoculated mice. Most likely, the expansion of this cell population occurs from cell division and not from infection since these diseased mice were found to harbor no virus replicating on mouse cells. It thus appears that the Duplan defective virus behaves as an oncogenic virus, stimulating the growth of its target cells.

Tumors induced by retroviruses or other agents in mice and in most other species are often clonal or oligoclonal, i.e., resulting from the growth of a single or a few cells which have gained a growth advantage over their neighboring cells. It is widely assumed that these cells have sustained numerous genetic changes which confer on them this growth advantage. We thus tested whether the cells infected with the defective Duplan virus in MAIDS mice were clonal (22). To do so, we attempted to detect virus-cell junction fragments with appropriate restriction endonucleases. In several, but not all, enlarged nodes of the diseased mice analyzed, we could detect novel fragments hybridizing with our probes (Fig. 7). These varied in intensity from node to node. These bands represented virus-cell junction fragments of newly integrated defective proviruses. In 10 out of 18 mice which have been analyzed to date, the pattern of provirus integration was different in each tested lymph node. In a few of the mice studied (4/18), the same integration pattern of proviruses was found in lymph nodes of the same mouse, indicating that the same clone of infected cells was proliferating in different anatomical sites (22, 23) (Fig. 7).

These results indicate that the infected target cells emerge as a clonal outgrowth later in the disease. Clonal expansion of these cells seems to be anatomically restricted in most instances and does not tend to migrate to other organs. In ~20% of diseased mice, however, one expanding clone acquires the ability to migrate and to grow in more than one anatomical site and behaves as a metastatic tumor. The clonality of these tumors suggests that either the infection of the target cells has been very inefficient and that only a few target cells within the body have been infected; alternatively, and we believe more likely, this clonality suggests that the defective virus is not

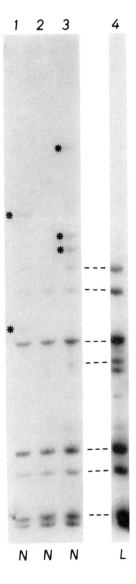

Figure 7. Detection of newly acquired defective proviruses in enlarged tissues of MAIDS mice. DNA from normal liver (L) or enlarged lymph nodes (N) was cleaved with *Sac*I (an enzyme which does not cleave the defective viral genome) and hybridized with ³²P-D30 probe as described (22). The presence of newly acquired fragments is indicated by an asterisk.

by itself sufficient to confer the full transformation potential to these target cells and that additional genetic events occurred in these infected cells. Therefore, the defective MAIDS virus behaves as an oncogenic virus and promotes the formation of a tumor of a specific cell type, yet to be determined. We propose that the immunodeficiency syndrome seen in these mice occurs as a consequence of the proliferation of these cells (22).

A Model of Pathogenesis of MAIDS

The knowledge accumulated on this disease appears, for the moment, sufficient to elaborate a model of pathogenicity (Fig. 8). The fact that helper-free stocks of defective MAIDS virus induce disease represents an important observation for our understanding of its pathogenesis. Following injection with helper-free stocks of defective MAIDS retrovirus, most likely several cells of different lineages expressing an ecotropic MuLV receptor get infected. These could be any cell type: fibroblastic, endothelial, muscular, hemato-poietic, or lymphoid cells, etc. We assume that a single cell type is the target for this virus and that the infection of this target cell population is essential and obligatory for the induction of the disease. The nature of these target cells is not known at the moment. Following integration of the defective genome into the various susceptible cells, the viral *gag* gene is expressed and the Pr60 *gag*/fusion protein is produced. In most of the cell types infected, the presence of the defective viral genome is benign and does not affect their metabolism. In the target cells, the defective viral genome induces prolifer-ation of each infected cell (polyclonal) and behaves as an oncovirus. We have now evidence that this proliferation occurs early after infection (within 5 to 10 days) (23). Proliferating cells are unlikely to be malignant at this stage.

The Pr60 *gag*/fusion protein, encoded by this virus, is most likely re-sponsible for this cell proliferation. It appears to be the only gene product encoded by the defective viral genome (21), and recent work with deleted

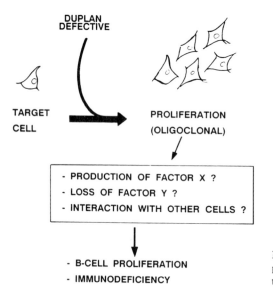

Figure 8. A model of MAIDS pathogenesis, as discussed in the text.

and mutated viral genomes indicates that this gene is essential for the induction of the disease (20). The mechanism by which such a protein could lead to cell proliferation is unclear at the moment and there are no previous experiments, to our knowledge, showing a direct role of *gag* proteins in growth deregulation. However, indirect evidence for such a role has been reported. During the construction of various chimeric MuLVs with several parental MuLVs exhibiting different leukemogenicity, it has been found that the *gag* region harbors minor determinants of pathogenicity (7, 19, 25, 41–43). Also the *gag* region of the avian osteopetrosis virus has been reported to harbor the major determinant of its pathogenicity (48). And recently, six amino acids from the amino terminus of *gag*, fused to the oncoprotein v-*erb*-B, have been shown to increase significantly the transformation potential of this protein (4). Therefore, it seems that, in some cell types, the *gag* proteins have very specific effects on cell metabolism. To date, the *gag* gene remains too poorly understood to predict its specific role(s) in each step of the virus cycle. For example, no function has ever been reported for the *gag* p12 protein, which was, for this reason, excluded from the current nomenclature of retroviral proteins (30). It is interesting that the most divergent sequences of the Duplan defective MAIDS Pr60*gag* reside within p12, suggesting it may harbor the determinant of pathogenicity for this virus.

When disease has progressed to the point that lymphadenopathy is important, the infected cells appear, in many nodes of most diseased mice, to represent a clonal or oligoclonal outgrowth (22). In our view, this observation suggests that secondary genetic events have occurred in these cells to allow them to gain a growth advantage over the other infected target cells. This clonal expansion is not too surprising and has been described in numerous types of tumors arising in different species. For example, the pre-B-cell expansion induced in mice by the defective Abelson virus (harboring the oncogene *abl*) is first polyclonal and then becomes clonal or oligoclonal as the disease progresses (13, 14). Proliferation of individual clones in anatomically restricted areas is frequent in MAIDS, and each node seems to be the site of the growth of a different clone (22). In some mice, the same clone seems to grow in more than one node and appears to arise from migrating cells. We postulate that these clones (able to migrate or to metastasize) have sustained additional genetic events conferring this phenotype. We do not know whether the tumor formed by these infected target cells is benign or malignant. The fact that Duplan's group has never been able to transplant diseased tissues (36) would suggest that these cells have not reached a full malignant phenotype, although this criterion may not be the best to assess the malignancy of some lymphoid or hematopoietic cells.

Since the infection of target cells is the very first event to occur following virus injection, and since expansion of this target cell population occurs early following infection and on a large scale (over 1,000-fold expansion), we proposed that proliferation of these target cells is involved in the disease and constitutes the primary and critical event in the disease (22). We proposed that the immunodeficiency seen in these mice develops as a consequence of this tumor growth, as a paraneoplastic syndrome. A prediction of this model would be that agents able to suppress the growth of the target cells would suppress MAIDS. We tested this hypothesis by using antineoplastic drugs, such as cyclophosphamide, a few days after infection or when MAIDS was already established (52). With both protocols, the drug was effective in preventing the appearance of the disease or in inducing its regression, as judged by the absence of lymph node and spleen enlargement and normal concanavalin A response of spleen T lymphocytes. Moreover, the number of infected cells (measured by the levels of defective proviral DNA or viral RNA) decreased significantly in these drug-treated, virus-inoculated mice. Paradoxically, an immunosuppressive drug (cyclophosphamide) was very effective in treating an immunodeficiency syndrome. These results are consistent with our model that growth of target cells is the primary inducing event of the immunodeficiency syndrome.

The model proposing that MAIDS is primarily induced by the proliferation of a specific target cell population predicts that any agents or genetic events able to stimulate the growth of these target cells, or to induce the putative factor that we hypothesized to be released by these proliferating cells (see below), will also induce MAIDS. Presumably the known characteristics of tumors of other tissues (such as multistep progression, long latency, oncogene activation, etc. [11, 54]) will also apply to this specific tumor. If the MAIDS model is relevant for other retrovirus-induced immunodeficiency syndromes, this prediction will also be valid for them. Therefore, these syndromes may occasionally arise in the absence of retrovirus infection as a consequence of a tumor growth arising spontaneously or following exposure to specific chemical or physical mutagens.

The mechanism by which proliferation of target cells leads to immunodeficiency is unknown, but we have proposed three hypotheses (22), as follows. (i) Infected target cells could interact with other cells of the immune system and lead to immunodeficiency. This model assumes that the target cells are expressing proteins at their surface which would be seen as different by the immune system and which would elicit some form of autoimmune disease. (ii) Infected target cells could have stopped secreting a factor essential to the immune system. This model appears unlikely because it assumes that the other uninfected cells of the target cell population will be unable to provide this factor. (iii) Infected target cells could secrete a factor which is detrimental to the immune system. Overproduction of this factor, which may be essential for a healthy immune system at lower concentration, leads to immune cell dysfunction directly or indirectly. We favor this hypothesis for reasons of simplicity, although no data favor any of these models at this point. MAIDS probably represents a cascade of many events influencing each other and involving many cell types of the immune system. For MAIDS to occur, most likely all these cells must be present and functional in the animal. It was therefore not too surprising to learn that T (39,57) and B (5) lymphoid cells are essential for its appearance.

The model proposed puts much emphasis on the target cells infected and their secreted factors (22). It is our feeling, however, that this cell population is a cell of a unique lineage and/or a cell at a unique stage of differentiation. It would be unlikely to be one of the frequent cell types transformed by other agents (viruses, chemicals, or irradiation) because these tumors would be accompanied by a severe immunodeficiency of the type seen in MAIDS. So, it is our prediction that agents generally used to induce cancer in mice are not transforming the cells which are targeted by the defective MAIDS virus, unless Pr60^{gag} is itself critical and induces specific factors in the target cells.

Many different types of cells have previously been reported to be proliferating in the enlarged organs of MAIDS mice. These involved oligoclonal expansion of T cells (26), polyclonal expansion of B cells (26, 28), oligoclonal nonneoplastic expansion of B cells (26), late oligoclonal malignant B-cell lymphoma (26, 45), expansion of macrophages (38), and proliferation and expansion of other non-T/non-B cells (28, 38). In most enlarged nodes, the infected target cells appear to represent only a proportion of the cells forming the tumor. The other cells in these enlarged organs appear to be reactive cells. The infected target cells that we have detected and which are stimulated to proliferate by the nonreplicating defective MAIDS virus may or may not represent one of the previously described cells in these tumors. The precise identification of the nature of the infected target cells with appropriate markers should clarify this issue.

Interestingly, the apparent underrepresentation of infected target tumor cells in MAIDS enlarged nodes appears similar to what is seen in Hodgkin's disease in humans. The cells which are thought to be malignant, the Reed-Sternberg cells, represent only a minority of the cells of the tumor mass, the other cells being normal or reactive cells, which have apparently migrated around tumor cells and which form the tumor mass (24).

Is MAIDS a Good Model for Human AIDS?

The immune system is very complex and involves interactions between several cell types producing and responding to a variety of factors. Each step of this complex network constitutes a potential target for various agents (chemicals, microorganisms, viruses, etc.) which may interfere with the immune homeostasis. Not all deregulation of the immune system, however, leads to what is known as an acquired immunodeficiency syndrome. For example, the nude and the SCID mouse mutants exhibit a severe immune dysfunction, but appear different from a MAIDS mouse and also lack some characteristics of the human AIDS, such as hypergammaglobulinemia. On the other hand, as a syndrome, AIDS is very difficult to define by a single parameter and is not always easily differentially diagnosed if the HIV status of the patient is not known, as was recently pointed out (9).

Indeed, several manifestations of AIDS, such as Kaposi's sarcoma and AIDS dementia syndrome, are not seen in all patients and therefore may not be strictly part of the syndrome. This problem raises the question of the criteria which should be used to compare human AIDS with the various acquired immunodeficiency syndromes seen in the animal models, and specifically with MAIDS.

The criteria of AIDS as a distinct human disease, independent of its etiology, may be summarized as (i) an early loss of T-cell functions, followed by a depletion of $CD4^+$ T-cells and (ii) a decrease of B-cell functions with B-cell proliferation (lymphadenopathy) and hypergammaglobulinemia. All the other manifestations (pneumonia by *Pneumocystis carinii*, diarrhea, terminal B-cell lymphoma, and various other bacterial, viral, or parasitic infections) appear to be secondary to the immune dysfunctions and cannot be considered as part of the definition of the primary disease. These few parameters of the AIDS definition are rather unspecific independently and are indeed found in other primary diseases. Together, however, they are slightly more specific.

If AIDS is an immune dysfunction of some specificity and if not all immune dysfunctions resemble AIDS, it follows that not all animal models of immune dysfunction will be good animal models for AIDS. There is no perfect animal model for AIDS available at present, although several retrovirus-induced AIDS-like diseases have been described (8, 12, 37). Among these, MAIDS appears to be a relatively good model for AIDS. As pointed out previously (37), the important manifestations of AIDS, and especially early AIDS, seem to be present in MAIDS: these include a rapid loss of helper T-cell functions preceding the loss of cytotoxic/suppressor T-cell functions, the loss of B-cell functions, polyclonal B-cell activation, and hypergammaglobulinemia. However, the B-cell proliferation is more extensive in MAIDS than in AIDS, and the depletion of $CD4^+$ cells seen in AIDS is not observed in MAIDS. Whether these differences represent epiphenomena related to species idiosynchrasia or essential differences reflecting distinct perturbations of the immune system is unknown.

The fact that MAIDS is not induced by a lentivirus but by a defective MuLV (oncovirus) does not necessarily disqualify it as a good model for

AIDS. Indeed, it is quite conceivable that HIV and the defective MAIDS virus could interfere (most likely through different molecular intermediates, because of their apparently distinct gene products) at the same or different steps of the same loop of the immune system, thus disturbing a common final pathway. For example, since the primary defect in MAIDS appears to be the proliferation of specific target cells, the same human target cell population could be stimulated to proliferate in AIDS by one of the HIV gene products. Alternatively, one of the HIV gene products, possibly the *tat* protein, could stimulate the production (without necessarily stimulating cell division) of the putative factor that we have postulated to be released by the proliferating MAIDS target cells and to be responsible for inducing immunodeficiency. It is very likely that both of these putative effects of HIV would have remained undetected to date.

Therefore, despite the fact that MAIDS is not induced by a lentivirus but by a defective MuLV, and despite its other apparent limitations, it still appears to be a valid model for AIDS and may in fact turn out to be very useful to understand the pathogenesis of AIDS. Indeed, the other animal models of AIDS, which are induced by a lentivirus, also show important differences as compared with AIDS (37). For example, simian AIDS induced by simian immunodeficiency virus is distinct from AIDS, as the disease has a very short latent period and as these animals are hypogammaglobulinemic and neutropenic and show no lymphadenopathy. Further comparisons of MAIDS and AIDS should tell the extent of their similarities and the significance of their differences.

Acknowledgments. The work done in our laboratory was supported by grants from the Medical Research Council of Canada and from the National Cancer Institute of Canada to P.J. We are grateful to D. Kay for reviewing the manuscript and to Marie Bernier for typing the manuscript. D.A. was the recipient of a fellowship from the National Cancer Institute of Canada.

REFERENCES

1. **Astier, T., B. Guillemain, F. Laigret, R. Mamoun, and J. F. Duplan.** 1982. Serological characterization of C-type retroviruses endogenous to the C57BL/6 mouse and isolated in tumours induced by radiation leukaemia virus (RadLV-Rs). *J. Gen. Virol.* **61:**55–63.
2. **Aziz, D. C., Z. Hanna, and P. Jolicoeur.** 1989. Severe immunodeficiency disease induced by a defective murine leukaemia virus. *Nature* (London) **338:**505–508.
3. **Bishop, J. M.** 1983. Cellular oncogenes and retroviruses. *Annu. Rev. Biochem.* **52:**301–354.
4. **Bruskin, A., J. Jackson, J. M. Bishop, B. J. McCarley, and R. C. Schatzman.** 1990. Six amino acids from the retroviral gene gag greatly enhance the transforming potential of the oncogene v-erb-B. *Oncogene* **5:**15–24.
5. **Cerny, A., A. W. Hügin, R. R. Hardy, K. Hayakawa, R. M. Zinkernagel, M. Makino, and H. C. Morse III.** 1990. B cells are required for induction of T cell abnormalities in a murine retrovirus-induced immunodeficiency syndrome. *J. Exp. Med.* **171:**315–320.

6. **Chattopadhyay, S. K., H. C. Morse III, M. Makino, S. K. Ruscetti, and H. W. Hartley.** 1989. Defective virus is associated with induction of murine retrovirus-induced immunodeficiency syndrome. *Proc. Natl. Acad. Sci. USA* **86:**3862–3866.

7. **DesGroseillers, L., and P. Jolicoeur.** 1984. Mapping the viral sequences conferring leukemogenicity and disease specificity in Moloney and amphotropic murine leukemia viruses. *J. Virol.* **52:**448–456.

8. **Desrosiers, R. C., and N. L. Letvin.** 1987. Animal models for acquired immunodeficiency syndrome. *Rev. Infect. Dis.* **9:**438–446.

9. **Duesberg, P. H.** 1989. Human immunodeficiency virus and acquired immunodeficiency syndrome: correlation but not causation. *Proc. Natl. Acad. Sci. USA* **86:**755–764.

10. **Fauci, A. S.** 1988. Immunodeficiency virus: infectivity and mechanisms of pathogenesis. *Science* **239:**617–622.

11. **Fearon, E. R., and B. Vogelstein.** 1990. A genetic model for colorectal tumorigenesis. *Cell* **61:**759–767.

12. **Gardner, M. B., and P. A. Luciw.** 1989. Animal models of AIDS. *FASEB J.* **3:**2593–2606.

13. **Green, P. L., D. A. Kaeler, and R. Risser.** 1987. Clonal dominance and progression in Abelson murine leukemia virus lymphomagenesis. *J. Virol.* **61:**2192–2197.

14. **Green, P. L., D. A. Kaeler, and R. Risser.** 1987. Cell transformation and tumor induction by Abelson murine leukemia virus in the absence of helper virus. *Proc. Natl. Acad. Sci. USA* **84:**5932–5936.

15. **Guillemain, B., T. Astier, R. Mamoun, and J. F. Duplan.** 1980. In vitro production and titration assays of B-tropic retroviruses isolated from C57BL mouse tumors induced by radiation leukemia virus (RadLV-Rs): effect of dexamethasone. *Intervirology* **13:**65–73.

16. **Guillemain, B., R. Mamoun, T. Astier, J. P. Portail, E. Legrand, and J. F. Duplan.** 1977. Studies on radiation leukemia virus released by a cell line derived from the leukemic spleen of a C57BL mouse, p. 297–309. *In* J. F. Duplan (ed.), *International Symposium on Radiation-Induced Leukemogenesis and Related Viruses.* Elsevier-North Holland, Amsterdam.

17. **Haas, M., and A. Meshorer.** 1979. Reticulum cell neoplasms induced in C57BL/6 mice by cultured virus grown in stromal hematopoietic cell lines. *JNCI* **63:**427–439.

18. **Haas, M., and T. Reshef.** 1980. Non-thymic malignant lymphomas induced in C57BL/6 mice by cloned dualtropic viruses isolated from hematopoietic stromal cell lines. *Eur. J. Cancer* **16:**909–917.

19. **Holland, C. A., J. W. Hartley, W. P. Rowe, and N. Hopkins.** 1985. At least four viral genes contribute to the leukemogenicity of murine retrovirus MCF 247 in AKR mice. *J. Virol.* **53:**158–165.

20. **Huang, M., Z. Hanna, and P. Jolicoeur.** Unpublished data.

21. **Huang, M., and P. Jolicoeur.** 1990. Characterization of the *gag*/fusion protein encoded by the defective Duplan retrovirus inducing murine acquired immunodeficiency syndrome. *J. Virol.* **64:**5764–5772.

22. **Huang, M., C. Simard, and P. Jolicoeur.** 1989. Immunodeficiency and clonal growth of target cells induced by helper-free defective retrovirus. *Science* **246:**1614–1617.

23. **Huang, M., C. Simard, and P. Jolicoeur.** Unpublished data.

24. **Jaffe, E. S.** 1989. The elusive Reed-Sternberg cell. *N. Engl. J. Med.* **320:**529–531.

25. **Jolicoeur, P., and L. DesGroseillers.** 1985. Neurotropic Cas-BR-E murine leukemia virus harbors several determinants of leukemogenicity mapping in different regions of the genome. *J. Virol.* **56:**639–643.

26. **Klinken, S. P., T. N. Fredrickson, J. W. Hartley, R. A. Yetter, and H. C. Morse III.** 1988. Evolution of B cell lineage lymphomas in mice with a retrovirus-induced immunodeficiency syndrome, MAIDS. *J. Immunol.* **140:**1123–1131.

27. **Latarget, R., and J. F. Duplan.** 1962. Experiment and discussion on leukaemogenesis by cell-free extracts of radiation-induced leukaemia in mice. *Int. J. Radiat. Biol.* **5:**339–344.

28. **Legrand, E., R. Dalculsi, and J. F. Duplan.** 1981. Characteristics of the cell populations involved in extra-thymic lymphosarcoma induced in C57BL/6 mice by RadLV-Rs. *Leuk. Res.* **5:**223–233.

29. **Legrand, E., B. Guillemain, R. Dalculsi, and F. Laigret.** 1982. Leukemogenic activity of B-ecotropic C-type retroviruses isolated from tumors induced by radiation leukemia virus (RadLV-Rs) in C57BL/6 mice. *Int. J. Cancer* **30:**241–247.

30. **Leis, J., D. Baltimore, J. M. Bishop, J. Coffin, E. Fleissner, S. P. Goff, S. Oroszlan, H. Robinson, A. M. Skalka, H. M. Temin, and V. Vogt.** 1988. Standardized and simplified nomenclature for proteins common to all retroviruses. *J. Virol.* **62:**1808–1809.

31. **Levy, J. A.** 1978. Xenotropic type C viruses. *Curr. Top. Microbiol. Immunol.* **79:**111–213.

32. **Lieberman, M., and H. S. Kaplan.** 1959. Leukemogenic activity of filtrates from radiation-induced lymphoid tumors of mice. *Science* **30:**387–388.

33. **Mamoun, R., B. Guillemain, T. Astier, J. P. Portail, E. Legrand, and J. F. Duplan.** 1978. Production et analyse d'un complexe viral dérivé d'un virus des radioleucoses de la souris C57BL. *Int. J. Cancer* **22:**98–105.

34. **Mann, R., R. C. Mulligan, and D. Baltimore.** 1983. Construction of a retrovirus packaging mutant and its use to produce helper-free defective retrovirus. *Cell* **33:**153–159.

35. **Marx, P. A., D. H. Maul, K. G. Osborn, N. W. Lerche, P. Moody, L. J. Lowenstine, R. V. Henrickson, L. O. Arthur, R. V. Gilden, M. Gravell, W. T. London, J. L. Sever, J. A. Levy, R. J. Munn, and M. B. Gardner.** 1984. Simian AIDS: isolation of a type D retrovirus and transmission of the disease. *Science* **223:**1083–1086.

36. **Mistry, P. B., and J. F. Duplan.** 1973. Propriétés biologiques d'un virus isolé d'une radiooleucémie C57BL. *Bull. Cancer* (Paris) **60:**287–300.

37. **Mosier, D. E.** 1986. Animal models for retrovirus-induced immunodeficiency disease. *Immunol. Invest.* **15:**233–261.

38. **Mosier, D. E., R. A. Yetter, and H. C. Morse III.** 1985. Retroviral induction of acute lymphoproliferative disease and profound immunosuppression in adult C57BL/6 mice. *J. Exp. Med.* **161:**766–784.

39. **Mosier, D. E., R. A. Yetter, and H. C. Morse III.** 1987. Functional T lymphocytes are required for a murine retrovirus-induced immunodeficiency disease (MAIDS). *J. Exp. Med.* **165:**1737–1742.

40. **Mullins, J. I., C. S. Chen, and E. A. Hoover.** 1986. Disease-specific and tissue specific production of unintegrated feline leukaemia virus variant DNA in feline AIDS. *Nature* (London) **319:**333–336.

41. **Oliff, A., M. D. McKinney, and O. Agranovsky.** 1985. Contribution of the *gag* and *pol* sequences to the leukemogenicity of Friend murine leukemia virus. *J. Virol.* **54:**864–868.

42. **Oliff, A., and S. Ruscetti.** 1983. A 2.4-kilobase-pair fragment of the Friend murine leukemia virus genome contains the sequences responsible for Friend murine leukemia virus-induced erythroleukemia. *J. Virol.* **46:**718–725.

43. **Oliff, A., K. Signorelli, and L. Collins.** 1984. The envelope gene and long terminal repeat sequences contribute to the pathogenic phenotype of helper-independent Friend viruses. *J. Virol.* **51:**788–794.

44. **Paquette, Y., D. G. Kay, E. Rassart, Y. Robitaille, and P. Jolicoeur.** 1990. Substitution of the U3 long terminal region of the neurotropic Cas-Br-E retrovirus affects its disease-inducing potential. *J. Virol.* **64:**3742–3752.

45. **Pattengale, P. K., C. R. Taylor, P. Twomey, S. Hill, J. Jonasson, T. Beardsley, and M. Haas.** 1982. Immunopathology of B-cell lymphomas induced in C57BL/6 mice by dualtropic murine leukemia virus (MuLV). *Am. J. Pathol.* **107:**362–377.

46. **Pedersen, N. C., E. Ho, M. L. Brown, and J. K. Yamamoto.** 1986. Isolation of a T-lymphotropic virus from domestic cats with an immunodeficiency-like syndrome. *Science* **235:**790–793.

47. **Poirier, Y., and P. Jolicoeur.** 1989. Distinct helper virus requirements for Abelson murine leukemia virus-induced pre-B- and T-cell lymphomas. *J. Virol.* **63**:2088–2098.

48. **Robinson, H. L., S. S. Reinsch, and P. R. Shank.** 1986. Sequences near the 5′ long terminal repeat of avian leukosis viruses determine the ability to induce osteopetrosis. *J. Virol.* **59**:45–49.

49. **Rosenberg, N., and D. Baltimore.** 1978. The effect of helper virus on Abelson virus-induced transformation of lymphoid cells. *J. Exp. Med.* **147**:1126–1141.

50. **Savard, P., L. DesGroseillers, E. Rassart, Y. Poirier, and P. Jolicoeur.** 1987. Important role of the long terminal repeat of the helper Moloney murine leukemia virus in Abelson virus-induced lymphoma. *J. Virol.* **61**:3266–3275.

51. **Scher, C. D.** 1978. Effect of pseudotype on Abelson virus and Kirsten sarcoma virus-induced leukemia. *J. Exp. Med.* **147**:1044–1053.

52. **Simard, C., and P. Jolicoeur.** 1991. The effect of anti-neoplastic drugs on murine acquired immunodeficiency syndrome. *Science* **251**:305–308.

53. **Towler, D. A., J. I. Gordon, S. P. Adams, and L. Glaser.** 1988. The biology and enzymology of eukaryotic protein acylation. *Annu. Rev. Biochem.* **57**:69–99.

54. **Weinberg, R. A.** 1989. Oncogenes, antioncogenes, and the molecular bases of multistep carcinogenesis. *Cancer Res.* **49**:3713–3721.

55. **Wolff, L., and S. Ruscetti.** 1985. Malignant transformation of erythroid cells in vivo by introduction of a nonreplicating retrovirus vector. *Science* **228**:1549–1552.

56. **Wong-Staal, F., and R. C. Gallo.** 1985. Human T-lymphotropic retroviruses. *Nature* (London) **317**:395–403.

57. **Yetter, R. A., R. M. L. Buller, J. S. Lee, K. L. Elkins, D. E. Mosier, T. N. Fredrickson, and H. C. Morse III.** 1988. CD4+ T cells are required for development of a murine retrovirus-induced immunodeficiency syndrome (MAIDS). *J. Exp. Med.* **168**:623–635.

Part III

ONCOGENESIS BY RETROVIRUSES

Viruses That Affect the Immune System
Edited by Hung Y. Fan et al.
© 1991 American Society for Microbiology, Washington, DC 20005

Chapter 9

Retrovirus Variation and Regulation of c-*rel*

Howard M. Temin, Mark Hannink, Wei-Shau Hu, and
Vinay K. Pathak

Retroviruses can play havoc with the immune system. For example, one species of avian retroviruses, the reticuloendotheliosis viruses, has members that cause immune deficiency (spleen necrosis virus) or leukemias (reticuloendotheliosis virus strain T[Rev-T]). In this brief article, we shall discuss how the error-prone replication of retroviruses provides genetic processes which enable retroviruses to exploit efficiently evolutionary opportunities like those posed by the evolution and development of the immune system. First, we discuss how the replication of retroviruses inevitably results in a population with a large number of genetic variants. Second, we discuss how the replication of retroviruses can transduce important cell genes away from their cellular controls, thus turning them into efficient cancer-causing genes.

RETROVIRUS VARIATION

Normal retrovirus replication is error prone (19); for example, random cloning and sequencing of retroviruses indicates a high degree of sequence variation (17). We developed a protocol to measure the rate of genetic change in a single cycle of retrovirus replication using retrovirus helper cells and vectors made from spleen necrosis virus, an oncoretrovirus (5). With this protocol and a retrovirus shuttle vector, we have now measured the rate of forward mutations in a single cycle of retrovirus replication (20, 21). With

Howard M. Temin, Mark Hannink, Wei-Shau Hu, and Vinay Pathak • McArdle Laboratory, University of Wisconsin, 1400 University Avenue, Madison, Wisconsin 53706.

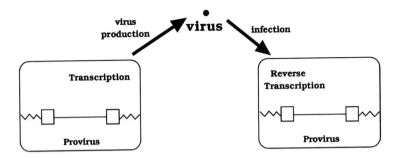

Figure 1. Single cycle of retrovirus replication. Proviruses include two long terminal repeats (open boxes) and coding sequences (straight line) integrated with cellular DNA (zig-zag lines). Two cells are shown.

other vectors and the same protocol, we have measured the rate of recombination in a single cycle of retrovirus replication (10).

Mutation

A single cycle of retrovirus replication goes from provirus to provirus and involves three rounds of nucleic acid polymerization: transcription from the provirus, reverse transcription of viral RNA, and plus-strand DNA synthesis (Fig. 1). Our retrovirus shuttle vector contained the *neo*[r] gene, which can be selected in mammalian cells with G418 and in bacterial cells with kanamycin, and a pBR322 origin of replication (Fig. 2, pVP212). Thus, this vector can replicate as a retrovirus in mammalian cells and as a plasmid in bacterial cells. The *lacZα* peptide was present as a reporter gene and could be scored for mutations in the proper bacterial cells. The mutations were characterized by DNA sequencing. Their frequencies gave rates of genetic changes.

Base pair substitutions occurred at a rate of 7×10^{-6} per nucleotide per

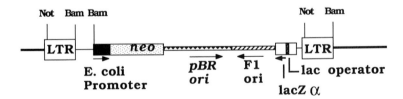

Figure 2. Retrovirus shuttle vector. The provirus of VP212 is shown. Cellular DNA is represented by heavy lines. Parts of the shuttle vector are labeled. The unlabeled light lines are other viral sequences. Arrows indicate direction of transcription or replication. Restriction enzyme cleavage sites are shown over the vector. Not, *Not*I; Bam, *Bam*HI.

cycle, frameshifts occurred at a rate of 1×10^{-6} per nucleotide per cycle, deletions occurred at a rate of 2×10^{-6} per nucleotide per cycle, and complex rearrangements occurred at a rate of 2×10^{-6} per nucleotide per cycle. The sequences of the complex rearrangements indicated that they occurred during plus-strand DNA synthesis and involved transfer, switching, or jumping of the growing point to a new template rather than to the neighboring nucleotide on the original template as normally occurs (Fig. 3). A similar process may underlie the formation of simple deletions and frameshifts (see also reference 22).

In addition, two other types of mutations were found at higher frequency. First, in the parental shuttle vector there was a direct repeat of about 110 bp containing a run of nine A residues. Direct sequencing indicated that one copy of the direct repeat was deleted at a rate of about 40% per replication cycle and that the length of the A run changed at a rate of about 15% per cycle. Another run of 10 T residues in the viral long terminal repeat also changed in length at a rate of 40% per cycle. Second, two proviruses were found with 15 base pair substitutions in a 930-bp region. We postulated that this high rate of mutation was the result of a mutant reverse transcriptase resulting from an epigenetic event, since other viruses from the same helper cell clone did not have this higher rate of mutation. Such hypermutant proviruses were found at a frequency of 10^{-4}, similar to the frequency of transcription/translation errors (24). This process is called hypermutation.

Recombination

A similar process of transfer, switching, or jumping of the growing point is probably involved in retrovirus recombination. To measure the rate of recombination in a single cycle of retrovirus replication, helper cells were infected with two different retrovirus vectors, each expressing two marker genes. One marker gene in each vector was inactivated by a frameshift mutation (Fig. 4) (10). Progeny proviruses were screened for proviruses expressing the two wild-type genes and were then characterized by restriction enzyme digestion. With two markers 1 kbp apart, 2% doubly resistant proviruses were found. The recombinants only appeared after the formation of heterodimeric virions.

To get the actual rate of recombination, two other factors must be considered. First, since the parental viruses differed by only 8 bp, there should be no bias in formation of virions, and heterodimers should occur about half of the time. Second, only one of the two possible recombinants was recovered. Together these factors indicate that the rate of recombination is four times higher than measured.

Further study using multiply marked parental viruses indicated that mul-

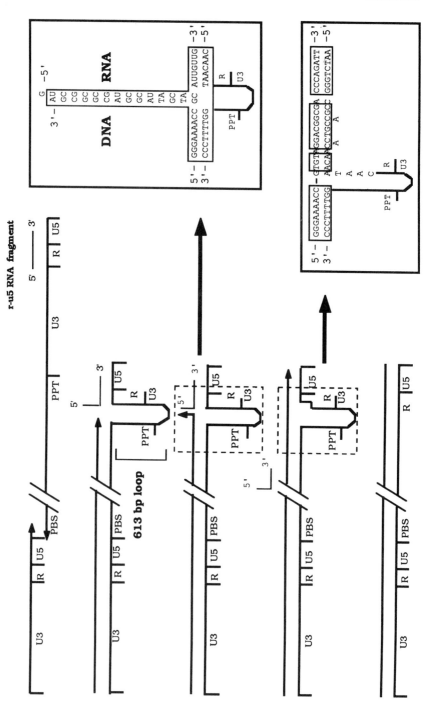

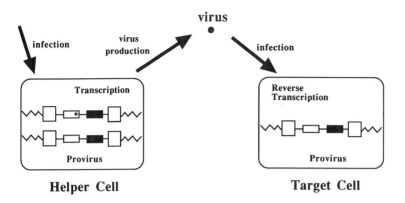

Figure 4. Recombination in a single cycle of retrovirus replication. Two retrovirus vectors are shown in a helper cell. Rectangles represent selectable genes. Asterisks represent frameshift mutations. Other symbols are as described in the legend of Fig 1.

tiple exchanges occurred during recombination (11). Moreover, the results of these experiments indicated that the exchanges took place both during reverse transcription and during plus-strand DNA synthesis.

Consequences of High Rate of Variation

As a result of these genetic processes, any retrovirus population will have members with many different nucleotide sequences. The rates of genetic change are so high that almost every individual retrovirus will have a different nucleotide sequence. Because of the high rate of recombination, the extent of this variation will be much larger than that resulting from mutation alone. Retroviruses thus differ quantitatively from other RNA viruses in the extent of sequence variation (4, 6).

Moreover, retroviruses, in contrast to other RNA viruses, have a second mode of replication, as a DNA provirus. As a DNA provirus, the retrovirus genome is replicated by an error-correcting DNA polymerase and thus is much more stable. This mode of replication allows "freezing" of more fit variants away from the error-prone retrovirus replication modes.

Figure 3. Complex rearrangement. A model for the occurrence of one of the complex rearrangements is shown. It is postulated that during plus-strand DNA synthesis the growing DNA molecule was elongated on a plus-strand RNA template that hybridized to complementary sequences in the minus-strand long terminal repeat, resulting in a 613-bp deletion. It is further postulated that the newly synthesized DNA continued using the minus-strand long terminal repeat as a template.

ORIGIN OF ONCOGENES

Another consequence of the existence of these two modes of retrovirus replication, as a virus or as a provirus, is that transduction of cellular sequences can occur as a result of transcription readthrough followed by missplicing and nonhomologous recombination (23). The additional mutations engendered during replication allow the later appearance of variants with selectable properties, such as tumor formation.

Rev-T

Rev-T is a highly oncogenic retrovirus that causes rapid, lethal lymphoid tumors in young galliform birds and transforms and immortalizes a small proportion of B-lymphoid cells from chicken spleen and bone marrow. Rev-T was formed from the recombination of a replication-competent retrovirus similar to reticuloendotheliosis virus strain A and nine exons of a turkey gene. There were also a deletion of *gag* and *pol* sequences from the parental retrovirus and numerous amino acid changes in the product of the turkey gene (see reference 2).

The product of the altered gene, v-*rel*, is 59 kDa, is cytoplasmic in transformed spleen cells and tumors, is nuclear in untransformed fibroblasts, transactivates certain promoters, is toxic to certain cells, and has an associated kinase activity. Although the difference in cellular location was thought to be responsible for the tissue specificity of transformation (8), later work with v-Rel proteins containing stronger nuclear targeting sequences showed that v-Rel proteins in either the nucleus or cytoplasm of spleen cells could transform (9). The v-Rel protein only transforms when expressed at a high level (see reference 18).

c-*rel*

Cloning of chicken cDNA and genomic DNA allowed construction of retrovirus vectors expressing the c-Rel protein. The c-Rel protein is 69 kDa, does not transform or cause tumors even when expressed at high levels, and is cytoplasmic in fibroblasts (3, 12, 15). c-Rel contains amino- and carboxy-terminal cytoplasmic targeting sequences, as well as the nuclear targeting sequences found in v-Rel.

The c-*rel* upstream sequences were also cloned and sequenced (13). There is not a TATA box promoter, but several Sp1 and NF-κB binding sites. The c-Rel (or v-Rel) protein down-regulates synthesis from the c-*rel* promoter by a factor of about 100. Thus, the expression of c-*rel* is normally tightly controlled. The weak promoter keeps c-Rel synthesis low, and when synthesis occurs, the presence of the c-Rel protein lowers synthesis again. Why the cell has this tight control on c-Rel is still not known, but the newly described relationship between *rel* and NF-κB (1, 7, 14, 16) suggests the control is part of the control of transcription factors.

Formation of Rev-T

Rel expression in Rev-T is no longer subject to the negative controls of c-*rel*, and thus the Rel protein is expressed at a high level from retrovirus regulatory sequences. Although the wild-type c-Rel protein itself does not cause cancer even when expressed at a high level (3, 15), the mutant Rel protein, v-Rel, does. Thus, the activation of c-*rel* to efficiently transform and immortalize was the result of substitution of strong control elements and the change in amino acids.

CONCLUSION

The formation of Rev-T can serve as a paradigm of retrovirus evolution. Retrovirus replication is extremely error prone and causes the formation of many different variants. When a variant with greater fitness appears, it forms a provirus which is selected directly or through its progeny and thus forms the basis of a new population. All retroviruses exist as a population of distinct individuals with a variety of nucleotide sequence variants. When conditions change or new evolutionary niches appear, a retrovirus variant able to take advantage is likely to be present.

Acknowledgments. We thank Gary Pulsinelli for useful comments on the manuscript. The research in our laboratory is supported by Public Health Service grants CA-22443 and CA-07175 from the National Cancer Institute. H.M.T. is an American Cancer Society Research Professor.

REFERENCES

1. **Ballard, D. W., W. H. Walker, S. Doerre, P. Sista, J. A. Molitor, E. P. Dixon, N. J. Peffer, M. Hannink, and W. C. Greene.** 1990. The v-*rel* oncogene encodes a κB enhancer binding protein that inhibits NF-κB function. *Cell* **63:**803–814.
2. **Bhat, G. V., and H. M. Temin.** 1990. Mutational analysis of v-*rel*, the oncogene of reticuloendotheliosis virus strain T. *Oncogene* **5:**625–634.
3. **Capobianco, A., D. L. Simmons, and T. D. Gilmore.** 1990. Cloning and expression of a chicken c-*rel* DNA: unlike p59^{sv-rel}, p68^{sc-rel} is a cytoplasmic protein in chicken embryo fibroblasts. *Oncogene* **5:**257–266.
4. **Domingo, E., and J. J. Holland.** 1988. High error rates, population equilibrium, and evolution of RNA regulatory systems, p. 3–36. *In* E. Domingo, J. J. Holland, and P. Ahlquist (ed.), *RNA Genetics*, vol. III, *Variability of RNA Genome*. CRC Press, Boca Raton, Fla.
5. **Dougherty, J. P., and H. M. Temin.** 1986. High mutation rate of a spleen necrosis virus-based retrovirus vector. *Mol. Cell. Biol.* **6:**4387–4395.

6. **Eigen, M., and C. K. Biebricher.** 1988. Sequence space and quasispecies distribution, p. 211–245. *In* E. Domingo, J. J. Holland, and P. Ahlquist (ed.), *RNA Virus Genetics*, vol. III, *Variability of RNA Genome.* CRC Press, Boca Raton, Fla.

7. **Ghosh, S., A. M. Gifford, L. R. Risiere, P. Tempst, G. P. Nolan, and D. Baltimore.** 1990. Cloning of the p50 DNA-binding subunit of NF-kB: homology to rel and dorsal. *Cell* **62:**1019–1029.

8. **Gilmore, T. D., and H. M. Temin.** 1986. Different localization of the product of the v-*rel* oncogene in chicken fibroblasts and spleen cells correlates with transformation by REV-T. *Cell* **44:**791–800.

9. **Gilmore, T. D., and H. M. Temin.** 1988. v-*rel* oncoproteins in the nucleus and in the cytoplasm transform chicken spleen cells. *J. Virol.* **62:**703–714.

10. **Hu, W.-S., and H. M. Temin.** 1990. Genetic consequences of packaging two RNA genomes in one retroviral particle: pseudodiploidy and high rate of genetic recombination. *Proc. Natl. Acad. Sci. USA* **87:**1556–1560.

11. **Hu, W.-S., and H. M. Temin.** 1990. Retrovirus reverse transcription and recombination. *Science* **250:**1227–1233.

12. **Hannink, M., and H. M. Temin.** 1989. Transactivation of gene expression by nuclear and cytoplasmic *rel* proteins. *Mol. Cell. Biol.* **9:**4323–4336.

13. **Hannink, M., and H. M. Temin.** 1990. Structure and autoregulation of the c-*rel* promoter. *Oncogene* **5:**1843–1850.

14. **Hannink, M., and H. M. Temin.** Unpublished data.

15. **Hannink, M., and H. M. Temin.** Unpublished data.

16. **Kieran, M., V. Blank, F. Logeat, J. Vandekerckhove, F. Lottspeich, O. Le Bail, M. B. Urgban, P. Kourilsky, P. A. Bauerle, and A. Isreal.** 1990. The DNA-binding subunit of NF-kB is identical to factor KBF1 and homologous to the rel oncogene product. *Cell* **62:**1007–1018.

17. **Meyerhans, A., R. Cheynier, J. Albert, M. Seth, S. Kwok, J. Sninsky, Morfeldt-Manson, B. Asjo, and S. Wain-Hobson.** 1989. Temporal fluctuations in HIV quasispecies in vivo are not reflected by sequential HIV isolations. *Cell* **58:**901–910.

18. **Miller, C. K., and H. M. Temin.** 1986. Insertion of several different DNAs in reticuloendotheliosis virus strain T suppresses transformation by reducing the amount of subgenomic mRNA. *J. Virol.* **58:**75–80.

19. **Pathak, V., W.-H. Hu, and H. M. Temin.** 1991. Hypermutations and other variations in retroviruses, p. 149–157. *In* E. J. Steele (ed.), *Somatic Hypermutation in V-Region.* CRC Press, Boca Raton, Fla.

20. **Pathak, V., and H. M. Temin.** 1990. Broad spectrum of *in vivo* forward mutations, hypermutations, and mutational hotspots in a retroviral shuttle vector after a single replication cycle: substitutions, frameshifts, and hypermutations. *Proc. Natl. Acad. Sci. USA* **87:**6019–6023.

21. **Pathak, V. K., and H. M. Temin.** 1990. Broad spectrum of *in vivo* forward mutations, hypermutations, and mutational hotspots in a retroviral shuttle vector after a single replication cycle: deletions and deletions with insertions. *Proc. Natl. Acad. Sci. USA* **87:**6024–6028.

22. **Roberts, J. D., B. D. Preston, L. A. Johnston, A. Soni, L. A. Loeb, and T. A. Kunkel.** 1989. Fidelity of two retroviral reverse transcriptases during DNA-dependent DNA synthesis in vitro. *Mol. Cell. Biol.* **9:**469–476.

23. **Temin, H. M.** 1988. Evolution of cancer genes as a mutation-driven process. *Cancer Res.* **48:**1697–1701.

24. **Yang, S.** Personal communication.

Viruses That Affect the Immune System
Edited by Hung Y. Fan et al.
© 1991 American Society for Microbiology, Washington, DC 20005

Chapter 10

Activated *abl* Genes Induce a Myeloproliferative Response in Mice

M. A. Kelliher, J. McLaughlin, O. N. Witte, and N. Rosenberg

Abelson murine leukemia virus (Ab-MLV), a murine retrovirus that expresses the v-*abl* oncogene, is classically associated with a rapid-onset pre-B-cell lymphoma called Abelson disease (16, 20). The protein tyrosine kinase activity of Abelson protein, the single product of v-*abl*, is required for tumor induction. A second form of activated *abl*, the P210 BCR/ABL protein associated with chronic myelogenous leukemia (CML), plays a central role in the dysregulated stem cell and myeloid lineage proliferation that characterizes this disease (2, 16, 20). Like Abelson protein, P210 is an activated protein tyrosine kinase. However, expression of P210 by a murine retrovirus is not sufficient for rapid tumor induction (12, 13).

The different efficiencies of transformation and the distinct disease spectra associated with Abelson protein and P210 BCR/ABL could reflect the differences in the structure of the two proteins. These differences include the unrelated nature of the *BCR*- and *gag*-derived sequences at the amino terminus of the two proteins and the presence of SH3 (*src* homologous 3) domain sequences in P210 (14, 17). The amino-terminal sequences are important in determining the subcellular localization of the proteins, while SH3 sequences play a central role in regulating the activity of normal ABL proteins (6, 8).

While it is tempting to conclude that structural differences between Abel-

M. A. Kelliher and N. Rosenberg • Department of Pathology, Department of Molecular Biology and Microbiology, and the Immunology Graduate Program, Tufts University School of Medicine, Boston, Massachusetts 02111.   **J. McLaughlin and O. N. Witte** • Department of Microbiology and Molecular Biology Institute, University of California at Los Angeles, Los Angeles, California 90024-1570.

son protein and P210 are central to the distinct diseases with which each is associated, other differences between Abelson disease and CML may be important. In particular, Abelson disease arises following expression of Abelson protein in pre-B lymphocytes (20). In contrast, CML is initiated by the translocation which generates the *BCR/ABL* gene fusion, an event that occurs in the hematopoietic stem cell (3, 16). To test the idea that the cell type in which activated *abl* proteins are expressed is important in controlling the disease process, we have introduced Abelson protein and P210 *BCR/ABL* into murine hematopoietic stem cell populations. When such populations are used to reconstitute lethally irradiated mice, about 50% of the animals develop a myeloproliferative syndrome. This disorder is characterized by elevated numbers of granulocytes in the peripheral blood and the presence of both transformed pre-B cells and macrophages.

CELLS AND VIRUSES

Ab-MLV-P160 (19) and an Ab-MLV-P90 strain (10), harvested from transformed NIH 3T3 cells, were used as a source of Ab-MLV. The JW-RX retrovirus, which expresses a *BCR/ABL* cDNA from Moloney murine leukemia virus promoter elements (12), was used for expression of P210 *BCR/ ABL*. Moloney murine leukemia virus, the helper virus present in Ab-MLV and JW-RX virus stocks, was prepared from chronically infected NIH 3T3 cells. Bone marrow cells were harvested from 9- to 12-week-old BALB/c mice that had been treated with a single dose (150 mg/kg of body weight) of 5-fluorouracil (5-FU) 6 days prior to sacrifice. The cells were mixed with undiluted virus and cultured in the presence of 20 U of recombinant interleukin-3 (gift of Dr. James Ihle) for 48 h. After this incubation, the cells were washed and injected into lethally irradiated (600 R followed by 300 R 3 h later) mice. Each animal received 10^5 cells. Normal animals injected with Ab-MLV within 48 h of birth were also evaluated.

GROWTH OF TUMOR CELLS AND BLOOD ANALYSIS

Animals were monitored for signs of tumor on a routine basis. Some Ab-MLV- and *BCR/ABL*-reconstituted mice were bled biweekly, and differential and total leukocyte counts were determined. When tumors developed, the animals were sacrificed and tissues were processed for histologic examination. Differential leukocyte counts and, in some cases, total leukocyte counts were performed.

Bone marrow cells were harvested from tumored animals and plated either in RPMI-1640, supplemented to contain 50 μM 2-mercaptoethanol and 20% heat-inactivated (56°C, 30 min) fetal calf serum, or minimal essential medium supplemented to contain 10% fetal calf serum. Expansion of tumor cell populations was monitored by microscopic evaluation of the cultures, and cells were subcultured when densities in excess of 2×10^6 cells per ml were reached. Alternatively, bone marrow cells were plated in soft agar medium as described previously (18). The number of colonies was assessed 10 to 12 days later. The morphology of cells in randomly selected colonies was assessed by staining with Wright-Giemsa stain. Representative colonies were transferred to liquid medium, and cell lines were derived (18).

The cell lines were analyzed by Wright-Giemsa, toluidine blue, myeloperoxidase, and alpha naphthyl esterase staining. The surface expression of antigens B220 (3), J11D (10), Fc receptor (25), and Mac-1 (24), Mac-2 (7), and Mac-3 (23) was assessed by staining with appropriate antibodies. The results were evaluated using a FACS IV instrument and compared with staining profiles obtained with appropriate negative control antibodies in all cases.

Ab-MLV-RECONSTITUTED MICE DEVELOP GRANULOCYTOSIS

A total of 19 mice reconstituted with Ab-MLV-infected, 5-FU-treated bone marrow were prepared. All developed tumors. However, because some animals died unexpectedly, only 14 mice were available for a complete pathology and cell culture workup. Gross examination revealed a picture consistent with typical Abelson disease in seven mice. Lymphadenopathy, tumors in the lower spinal region, and slight splenomegaly were prominent features. Analysis of the peripheral blood picture in the animals examined revealed that all of them had a pattern indistinguishable from that observed in mock-reconstituted animals or in animals that developed Abelson diseases following inoculation of virus as neonates (Table 1). In these cases, the blood picture did not differ significantly from that observed in normal mice of a similar age (21). This picture is consistent with the fact that Abelson disease is a lymphoma and malignant cells are not found in the circulation (15, 22).

Examination of the remaining seven mice revealed a unique disease pattern. These animals displayed marked splenomegaly and the presence of multiple 3- to 5-mm nodules of macrophage tumor cells. Slight lymphadenopathy and no spinal tumors were observed in these mice. The peripheral blood cell picture in those animals examined did not resemble that found in control animals or animals afflicted with Abelson disease. The percentage of lymphocytes was reduced 5- to 10-fold, and the percentage of granulocytic cells was elevated two- to threefold (Table 2). Such a pattern, coupled with the presence of prominent macrophage tumors, is consistent with a myelo-

Table 1. Peripheral Blood Analysis of v-*abl* Mice with Pre-B-Cell Lymphoma

Treatment[a]	Virus	Animal no.	Differential leukocyte count[b]				
			L	M	N	E	B
5-FU R	Yes	1	62	17	20	1	<1
		2	72	11	16	1	<1
		3	56	5	36	2	<1
5-FU R	No	1	51	6	39	<1	<1
		2	71	8	19	2	<1
None	Yes	1	60	3	35	3	<1
		2	71	5	23	1	<1

[a]5-FU R, animals reconstituted with bone marrow from 5-FU-treated bone marrow; None, unmanipulated mice injected with Ab-MLV within 48 h of birth.
[b]Differential leukocyte counts were determined using blood smears obtained at sacrifice that were stained with Wright-Giemsa stain. The values given are percentages. L, lymphocytes; M, monocytes; N, neutrophils; E, eosinophils; B, basophils.

proliferative syndrome, a disease that has not been reported following Ab-MLV infection (16, 20).

TRANSFORMED MACROPHAGE AND PRE-B-CELL LINES FROM MICE WITH MYELOPROLIFERATIVE DISEASE

The ability of bone marrow pre-B cells to grow in soft agar medium correlates well with Ab-MLV transformation (18). To determine which animals contained Ab-MLV-transformed cells, the bone marrow from 13 Ab-MLV-reconstituted mice and 2 normal mice injected with Ab-MLV as neonates was plated in agar. Colonies of transformed pre-B cells were observed in bone marrow from five of the seven reconstituted mice classified as having pre-B lymphoma and from three of the six reconstituted mice classified as

Table 2. Peripheral Blood Analysis of v-*abl* Mice with Myelomonocytic Disease[a]

Treatment	Virus	Animal no.	Differential leukocyte count				
			L	M	N	E	B
5-FU R	Yes	1	4	8	87	1	<1
		2	15	3	82	<1	<1
5-FU R	No	1	51	6	39	<1	<1
		2	71	8	19	2	<1

[a]The samples were processed as described in the footnotes of Table 1. For convenience of the reader, the values for the control mice shown in Table 1 are repeated here. For abbreviations, see Table 1, footnote b.

having the myeloproliferative disease (Table 3). The number of colonies obtained varied from animal to animal, a feature that may reflect the tumor load of the animals. The ability to isolate transformed pre-B cells from mice with myeloproliferative disease may indicate that more than one cell type is infected by the virus. Alternatively, this result may indicate that progenitor cells capable of giving rise to both macrophages and pre-B cells are responsible for the tumor. Analysis of proviral integrations (data not shown) suggests that the latter explanation is correct.

In contrast to the Ab-MLV-reconstituted mice and normal mice with pre-B-cell disease, bone marrow from all of the animals with the myeloproliferative syndrome gave rise to cells with macrophage morphology. As with the pre-B-cell colonies, the frequency of this type of colony varied from animal to animal (Table 3). In five cases, both macrophage and pre-B-cell lines were obtained from an individual animal. The ability to grow macrophages from the tumored mice in vitro, under conditions where large colonies of normal macrophages are not observed, suggests that the cells have undergone malignant transformation. Consistent with this idea, analysis of cellular extracts revealed the presence of Abelson protein in all colonies examined (data not shown).

Table 3. Presence of Transformed Cells in v-*abl* Mice[a]

Treatment	Diagnosis	Animal no.	Colonies/10^6 NBC		Mass cultures	
			L	M	L	M
5-FU R	Pre-B lymphoma	1	2	<0.5	4/8	0/8
		2	215	<0.5	NT	NT
		3	<0.5	<0.5	1/4	0/4
		4	168	<0.5	3/4	1/4
		5	<0.5	<0.5	4/4	0/4
		6	3	<0.5	2/2	0/2
		7	111	<0.5	NT	NT
5-FU R	Myelomonocytic leukemia	1	<0.5	<0.5	0/4	1/4
		2	2	<0.5	3/4	1/4
		3	<0.5	126	2/4	4/4
		4	<0.5	30	1/4	4/4
		5	9	3	NT	NT
		6	9	3	NT	NT
None	Pre-B lymphoma	1	82	<0.5	2/2	0/2
		2	114	<0.5	1/1	0/1

[a]Animals and bone marrow were treated as described in the text. Animals listed under Treatment as None were injected with Ab-MLV within 48 h of birth. For the agar assays, the number of colonies per 10^6 nucleated blood cells (NBC) is shown. For mass cultures, the numerator represents the number of cultures giving rise to cell lines and the denominator represents the total number of cultures observed. L, lymphoid; M, myeloid; NT, not tested.

The cell lines derived from 10 of the Ab-MLV-reconstituted mice were characterized for expression of a variety of cell surface markers associated with lymphoid and myeloid lineage cells. The staining pattern observed for the pre-B-cell lines derived from five of six mice was identical to that observed for typical Ab-MLV-transformed pre-B cells (11). The cells were uniformly positive for B220 and J11D and did not express Mac-1, Mac-2, or Mac-3 (Table 4). Approximately half of the lymphoid cells derived from one mouse were J11D positive and predominantly B220 negative. Although these cells have rearranged immunoglobulin heavy chain genes (data not shown), further analyses are needed to determine the relationship of these cells to other Ab-MLV-transformed pre-B cells. Consistent with the lymphoid nature of these cells, they did not stain with toluidine blue, alpha naphthyl esterase, or myeloperoxidase.

In contrast to the lymphoid cell lines, the macrophage cell lines were negative for J11D and B220 (Table 4). Most of these cells expressed Mac-1 and Mac-2, but none expressed Mac-3, a determinant most commonly associated with activated macrophage cells. The cells were negative for toluidine blue and myeloperoxidase stain and positive for alpha naphthyl esterase stain. Thus, these cells displayed the phenotypic pattern expected for macrophage cells. Preliminary results indicate that at least some of the cell lines have limited phagocytic capacity.

MYELOMONOCYTIC DISEASE IN *BCR/ABL*-RECONSTITUTED MICE

A total of 12 mice reconstituted with JW-RX-infected 5-FU marrow were prepared. The animals were bled biweekly beginning at day 30 post reconstitution, and differential and total leukocyte counts were performed. Even at this early time point, several animals displayed elevated total leukocyte counts, and the distribution of leukocytes was skewed in favor of granulocytes (Table 5 and data not shown). By the time obvious tumors developed, four of the mice had a peripheral blood picture consistent with myeloproliferative disease (Table 6). The other four mice examined in detail had a blood picture in the normal range. Consistent with the peripheral blood picture, autopsy examination revealed that animals with elevated numbers of myeloid lineage cells in the peripheral blood had myeloproliferative disease while those with a normal blood picture had pre-B-cell lymphomas.

The bone marrow cells from *BCR/ABL*-reconstituted mice were plated in liquid cultures and in the soft agar assay (18). Similar to the Ab-MLV-reconstituted mice, the bone marrow from three of the four animals with pre-B-cell disease gave rise to pre-B cell lines (Table 7). However, colonies

Table 4. Expression of Differentiation Parameters by Cell Lines from v-*abl*-Reconstituted Mice[a]

Mouse no.	No. of cell lines	Phenotypic characteristic[a]								
		B220	J11D	FcR	Mac-1	Mac-2	Mac-3	NSE	MYP	TB
1	6	+	+	NT	–	NT	NT	–	–	–
2	1	–	–	+	+	NT	NT	+	–	–
	1	–	–	+	–	NT	NT	+	–	–
3	1	–	+	+	–	NT	NT	–	–	–
	1	–	–	+	–	NT	NT	+	–	+
4	3	+	+	NT	–	NT	NT	+	–	–
	1	–	–	+	+	NT	NT	+	–	–
5	6	+	+	+	–	NT	NT	–	–	–
	11	–	+	+	–	NT	NT	–	–	–
6	22	+	+	NT	–	–	–	–	–	–
7	2	+	+	NT	–	–	–	–	–	–
	1	–	–	NT	–	–	–	+	–	+
	15	–	–	NT	+	+	–	+	–	–
8	2	+	+	NT	–	–	–	–	–	–
9	1	+	+	NT	–	–	–	–	–	–
	2	–	–	NT	+	+	–	+	–	–
10	2	+	+	NT	–	NT	NT	–	–	–

[a]Cell lines were analyzed as described in the text. +, >95% of the cells expressed the marker; –, <5% of the cells expressed the marker. FcR, Fc receptor; NSE, alpha naphthyl esterase; MYP, myeloperoxidase; TB, toluidine blue; NT, not tested.

Table 5. Total Leukocyte Counts of
BCR/ABL-Reconstituted Mice

Mouse no.	Total leukocyte count[a] at days postreconstitution		
	30	50	70
1	10,000	4,700	2,300
2	8,200	3,550	12,300
3	13,400	1,600	NA
4	10,600	5,700	3,800
5	11,400	2,500	2,050
6	47,600	16,750	84,400
7	6,200	1,900	12,400
8	12,800	700	2,800
9	1,950	1,350	21,000
10	1,350	1,500	4,400
11	1,600	1,950	1,950
12	2,050	1,500	6,200
Mock	6,400	1,950	1,950

[a]Total leukocyte counts were determined by counting appropriately diluted cells in a hemacytometer. The example shown for the mock animal is representative of values obtained from five animals reconstituted with mock-infected bone marrow from 5-FU-treated mice. NA, animal not available for examination.

were not observed in all the agar assays. Further, the colonies that were observed were smaller, reflecting the presence of fewer cells. These differences may stem from the reduced transforming potency of P210 *BCR/ABL*, a feature well documented using several in vitro systems (5, 12, 13). In contrast

Table 6. Peripheral Blood Analysis of *BCR/ABL* Mice at Sacrifice[a]

Diagnosis	Animal no.	Differential cell count					Total leukocyte count
		L	M	N	E	B	
Myelomonocytic leukemia	4	12	66	22	<1	<1	25,000
	12	60	10	31	<1	<1	12,300
Granulocytic leukemia	9	16	21	63	<1	<1	48,000
	6	2	<1	98	<1	<1	85,000
Pre-B lymphoma	1	70	5	24	1	<1	3,000
	2	80	3	14	3	<1	4,150
	8	52	<1	48	<1	<1	6,200
	11	67	6	25	2	<1	3,000

[a]Peripheral blood analysis was conducted as described in the footnotes of Table 1 and Table 4. For abbreviations, see Table 1, footnote *b*.

Table 7. Presence of Transformed Cells in *BCR/ABL*-Reconstituted Mice[a]

Diagnosis	Animal no.	Colonies/10^6 NBC		Mass cultures	
		L	M	L	M
Pre-B lymphoma	1	<0.5	<0.5	0/4	0/4
	2	<0.5	<0.5	2/4	0/4
	8	44	<0.5	2/4	0/4
	11	<0.5	<0.5	1/4	0/4
Myelomonocytic leukemia	4	<0.5	<0.5	0/4	0/4
	12	<0.5	<0.5	0/4	0/4
Granulocytic leukemia	6	<0.5	<0.5	0/4	0/4
	9	<0.5	<0.5	0/4	0/4

[a]Cells were grown as described in the text. For agar cultures, the values given represent the number of colonies per 10^6 nucleated blood cells (NBC). For the mass cultures, the numerator represents the number of cultures giving rise to cell lines and the denominator represents the total number of cultures examined. L, lymphoid; M, myeloid.

to the Ab-MLV-reconstituted mice, macrophage colonies were not obtained from the mice with myeloproliferative disease. The failure to isolate such cells may reflect the absence of large macrophage tumors in these mice.

Although subtle differences, described in detail elsewhere (9), distinguish Ab-MLV- and *BCR/ABL*-reconstituted mice, the general pattern observed suggests that introduction of an activated *abl* gene into an appropriate target cell is the major requirement for induction of myeloproliferative disease. A disease pattern similar to the myeloproliferative disease observed in these studies has been observed by Daley and co-workers (5) using an identical protocol. These investigators used only *BCR/ABL* expressed from a myeloid-specific promoter. However, our ability to induce the myeloid disease with both Ab-MLV and *BCR/ABL* expressed from a less-specific promoter suggests that neither features unique to *BCR/ABL* nor lineage-specific expression is central to the disease process.

The myeloproliferative syndrome observed in the reconstituted mice shares many features with the chronic phase of CML. The ability to induce this disease in a manipulable animal system will prepare the way for detailed assessment of the role of activated *abl* genes in controlling proliferation of hematopoietic progenitors. Further modification of the protocol may also allow study of the features controlling the transition from the chronic to the blast crisis phase of CML.

Acknowledgments. This work was supported by grants CA 24220 and CA 33771 to N.R. and grants CA 27507 and CA 32737 to O.N.W. from the National Institutes of Health. O.N.W. is an Investigator of the Howard Hughes Medical Institute. We are grateful to Carol Crookshank for her assistance in preparing this manuscript.

REFERENCES

1. **Bruce, J., S. W. Symington, T. J. McKearn, and J. Sprent.** 1981. A monoclonal antibody discriminating between subsets of T and B cells. *J. Immunol.* **127**:2496–2501.

2. **Champlin, R. E., and D. W. Golde.** 1985. Chronic myelogenous leukemia: recent advances. *Blood* **65**:1039–1047.

3. **Coffman, R., and I. L. Weissman.** 1981. B220: a B cell specific member of the T200 glycoprotein family. *Nature* (London) **289**:681–683.

4. **Daley, G. Q., J. McLaughlin, O. N. Witte, and D. Baltimore.** 1987. The CML-specific P210 *bcr/abl*, unlike v-*abl*, does not transform NIH/3T3 fibroblasts. *Science* **237**:532–537.

5. **Daley, G. Q., R. Van Etten, and D. Baltimore.** 1990. Induction of chronic myelogenous leukemia in mice by the P210 BCR/ABL gene of the Philadelphia chromosome. *Science* **247**:824–830.

6. **Franz, W. M., P. Berger, and J. Y. J. Wang.** 1989. Deletion of an N-terminal regulatory region of the c-*abl* tyrosine kinase activates its oncogenic potential. *EMBO J.* **8**:137–147.

7. **Ho, M.-K., and T. Springer.** 1981. Mac-2, a novel 32,000 M_r mouse macrophage subpopulation-specific antigen defined by monoclonal antibodies. *J. Immunol.* **128**:1221–1228.

8. **Jackson, P., and D. Baltimore.** 1989. N-terminal mutations activate the leukemogenic potential of the myristylated form of c-*abl. EMBO J.* **8**:449–456.

9. **Kelliher, M. A., J. McLaughlin, O. N. Witte, and N. Rosenberg.** 1990. Induction of a chronic myelogenous leukemia-like syndrome in mice with v-*abl* and BCR/ABL. *Proc. Natl. Acad. Sci. USA* **87**:6649–6653.

10. **Kelliher, M. A., and N. Rosenberg.** Unpublished data.

11. **McKearn, J. P., and N. Rosenberg.** 1985. Mapping cell surface antigens on mouse pre-B cell lines. *Eur. J. Immunol.* **15**:295–298.

12. **McLaughlin, J., E. Chianese, and O. N. Witte.** 1987. In vitro transformation of immature hematopoietic cells by the P210 *bcr/abl* oncogene product of the Philadelphia chromosome. *Proc. Natl. Acad. Sci. USA* **84**:6558–6562.

13. **McLaughlin, J., E. Chianese, and O. N. Witte.** 1989. Alternative forms of the *BCR/ABL* oncogene have quantitatively different potencies for stimulation of immature lymphoid cells. *Mol. Cell. Biol.* **9**:1866–1874.

14. **Mes-Masson, A.-M., J. McLaughlin, G. Q. Daley, M. Paskind, and O. N. Witte.** 1986. Overlapping cDNA clones define the complete coding region for the P210BCR/ABL gene product associated with chronic myelogenous leukemia cells containing the Philadelphia chromosome. *Proc. Natl. Acad. Sci. USA* **83**:9768–9772.

15. **Rabstein, L. S., A. F. Gazdar, H. C. Chopra, and H. T. Abelson.** 1971. Early morphological changes associated with infection by a murine nonthymic lymphatic tumor virus. *J. Natl. Cancer Inst.* **46**:481–491.

16. **Ramakrishnan, L., and N. Rosenberg.** 1989. *abl* genes. *Biochim. Biophys. Acta* **989**:209–224.

17. **Reddy, E. P., M. J. Smith, and A. Srinivasan.** 1983. Abelson murine leukemia virus genome: structural similarity of its transforming gene product to other *onc* gene products with tyrosine-specific kinase activity. *Proc. Natl. Acad. Sci. USA* **80**:3623–3627.

18. **Rosenberg, N., and D. Baltimore.** 1976. A quantitative assay for transformation of bone marrow cells by Abelson murine leukemia virus. *J. Exp. Med.* **143**:1453–1463.

19. **Rosenberg, N., and O. N. Witte.** 1980. Abelson murine leukemia virus mutants with alterations in the virus-specific P120 molecule. *J. Virol.* **33**:340–348.

20. **Rosenberg, N., and O. N. Witte.** 1988. The viral and cellular forms of the Abelson (*abl*) oncogene. *Adv. Virus Res.* **35**:39–81.

21. **Schalm, O. W., N. C. Jain, and E. J. Carroll.** 1975. *Veterinary Immunology*, 3rd ed., p. 219–283. Lea & Febiger, Philadelphia.

22. **Siegler, R., S. Zajdel, and I. Lane.** 1972. Pathogenesis of Abelson-virus-induced murine leukemia. *J. Natl. Cancer Inst.* **48:**189–218.
23. **Springer, T.** 1981. Monoclonal antibody analysis of complex biological systems. *J. Biol. Chem.* **256:**3833–3839.
24. **Springer, T., G. Galfre, D. S. Secher, and C. Milstein.** 1979. Mac-1: a macrophage differentiation antigen identified by monoclonal antibodies. *Eur. J. Immunol.* **9:**301–306.
25. **Unkeless, J.** 1979. Characterization of a monoclonal antibody directed against mouse macrophage and lymphocyte Fc receptors. *J. Exp. Med.* **150:**580–596.

Viruses That Affect the Immune System
Edited by Hung Y. Fan et al.
© 1991 American Society for Microbiology, Washington, DC 20005

Chapter 11

Leukemogenesis by Moloney Murine Leukemia Virus

Hung Fan, B. Kay Brightman, Brian R. Davis, and Qi-Xiang Li

The focus of this chapter is leukemogenesis by Moloney murine leukemia virus (M-MuLV). In particular, events that M-MuLV induces in the animal at early (preleukemic) times will be highlighted. While some of these events might be specific to M-MuLV-induced disease, it is likely that other MuLVs and maybe other nonacute retroviruses induce similar processes.

MOLECULAR AND VIRAL EVENTS IN MuLV LEUKEMOGENESIS

Murine leukemia viruses are classical replication-competent retroviruses that lack oncogenes. They induce leukemias or other hematopoietic neoplasms with long latency. The long disease time course might reflect (i) a multiple-step process for leukemogenesis, in which the virus induces one or more changes in a sequential process, or (ii) a stochastic process, in which a single event is of fundamental importance but occurs randomly at very low frequency.

In the past 10 years, several important principles have been identified about molecular and virological events involved in leukemogenesis by non-acute retroviruses, as follows.

Hung Fan, B. Kay Brightman, and Qi-Xiang Li • Department of Molecular Biology and Biochemistry and Cancer Research Institute, University of California, Irvine, California 92717. **Brian R. Davis** • Medical Research Institute, 2200 Webster Street, San Francisco, California 94115.

LTR Activation of Proto-Oncogenes

The process of activation of proto-oncogenes by the long terminal repeat (LTR) was first elucidated for avian leukosis virus-induced B-cell lymphomas by Hayward et al. (21). End-stage tumors all show proviruses integrated next to the c-*myc* proto-oncogene, which results in high-level readthrough transcription from the downstream avian leukosis virus LTR into the c-*myc* sequences ("promoter insertion," see Fig. 1). Subsequently, this mechanism has been generalized in that the enhancers in the LTRs can activate transcription of adjacent proto-oncogenes, even when classical promoter insertion does not take place ("LTR activation," see Fig. 1) (32, 33, 40).

The hallmark for LTR activation is that multiple independent tumors show provirus insertion at common integration sites. For M-MuLV, sev-

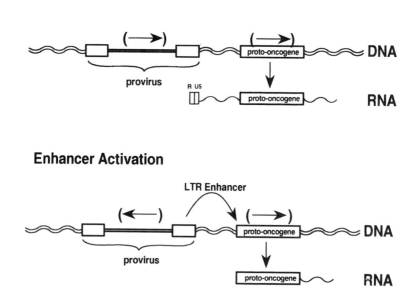

Figure 1. Proto-oncogene activation during nonacute retrovirus leukemogenesis. The promoter insertion mechanism is illustrated at the top of the diagram. In this case, transcription initiates in the downstream LTR of an inserted provirus and proceeds into proto-oncogene coding sequences. The result is overexpression of the proto-oncogene (as a hybrid transcript) under control of the retroviral LTR. The enhancer activation mechanism is illustrated at the bottom of the diagram. In this case, the LTR enhancer activates transcription from the proto-oncogene via its own promoter.

eral common integration sites are utilized in an either-or fashion, including c-*myc* (27, 37) (and several distinct but closely linked loci such as *pvt-/ Mlvi-1* and *Mlvi-4* [16, 22]), *pim-1* (8), *pim-2* (4), *Mlvi-2* and *-3* (41), and *Dsi-1* (43).

While retrovirus integration generally takes place at multiple (virtually random) sites within the host genome, MuLV-induced tumors are mono- or oligoclonal in nature, indicating expansion of an individual infected tumor cell, presumably due to an LTR-activated proto-oncogene. Given the low probability of proviral insertion at any given site on a random basis, this supports a stochastic basis for the long latency of nonacute retroviral disease.

It should be noted that it is unclear whether LTR activation of proto-oncogenes occurs in all nonacute retrovirus-induced tumors. For some viruses, the great majority of tumors show this process, e.g., c-*myc* for avian leukosis virus-induced B-lymphoma (21) and the proto-oncogenes activated by M-MuLV discussed above (15). However, Mucenski et al. (31) surveyed tumors induced by endogenous retroviruses in AKXD recombinant inbred lines of mice and found that only 22% showed proviral integrations at previously identified common insertion sites. We also studied T-lymphoid and B-lymphoid tumors induced by enhancer variants of M-MuLV (15, 17), and in some cases, the tumors showed very low incidence of proviral insertion at sites used by wild-type M-MuLV. This might indicate that other as yet unidentified proto-oncogenes were insertionally activated in these tumors. Alternatively, other virus-driven mechanisms may also cause tumor formation.

Tissue-Specific LTR Enhancers

Nonacute retroviruses generally induce very specific neoplasms. For instance, M-MuLV induces T-lymphoblastic lymphoma when inoculated into neonatal mice, while Friend MuLV induces erythroid leukemia. A number of laboratories have mapped the regions of the genome that control nonacute retroviral disease specificity and potency (7, 12, 24). The transcriptional enhancers in the LTRs were found to be the major disease determinant. Moreover, disease specificity was correlated with the strength of the LTR enhancers in the target cell type (36, 45). For instance, M-MuLV has enhancers that function efficiently in T-lymphoid cells, while Friend MuLV has enhancers that function well in erythroid cells. This is logical, given LTR activation of proto-oncogenes. Since the enhancers are responsible for activating the adjacent cell proto-oncogenes, they would do this most efficiently in cells where they are highly active.

MCF Recombinants

For MuLVs, leukemogenesis in mice is generally accompanied by appearance of *env* gene recombinants (see Coffin et al., this volume). These recombinants, called "mink cell focus-inducing" (MCF) viruses (20), result from recombination between the initial MuLV and endogenous retroviruses in the mouse genome. MCF *env* sequences bind to a different cellular receptor than do ecotropic MuLVs and confer expanded host range to both murine and nonmurine cells.

Several lines of evidence suggest that MCF recombinants play important roles in the leukemogenic process. Most MuLV-induced tumors are infected with MCF recombinants, indicating a positive selection for MCF recombinants during tumorigenesis (35, 42). In mouse strains genetically resistant to MCF formation (Rmcf), the time course of MuLV leukemogenesis is significantly extended (34). In spontaneous leukemia of AKR mice (arising from genetically transmitted Akv MuLV), AKR MCF viruses will accelerate the disease time course (20).

The exact role(s) of MCF recombinants in MuLV leukemogenesis is somewhat unclear. Some possibilities include (i) allowing additional rounds of infection in cells already infected with ecotropic MuLV by overcoming superinfection resistance; (ii) targeting infection to cells with high MCF receptor densities, (iii) growth-stimulatory effects of MCF envelope glycoprotein, and (iv) lytic effects on cells such as thymic stroma (see below).

Depending on the MuLV under study, pathogenic MCF recombinants may represent single or multiple recombinations. In spontaneous AKR leukemia, the genetically transmitted Akv MuLV undergoes three recombinations to generate the pathogenic MCF recombinant (14, 39), in which LTR sequences and MCF *env* glycoprotein have been acquired from two different endogenous viruses. The LTR recombination is necessary, because Akv MuLV enhancers are poorly active in T-lymphoid cells (45); thus an MCF recombinant with an Akv MuLV LTR would not efficiently activate proto-oncogenes in those cells. In contrast, M-MuLV MCFs generally only show changes in the *env* region (3), presumably because the M-MuLV enhancers are very active in T-lymphoid cells (36).

While MCF recombinants play an important role(s) in MuLV leukemogenesis in mice, they may not be absolutely required. For instance, Rmcf mice, which are genetically resistant to MCF formation, develop myeloid leukemia after inoculation with Friend MuLV without generating MCFs, but with a longer latency (34). Moreover, M-MuLV is also leukemogenic in rats, which lack endogenous MuLV proviruses and cannot generate MCFs (37, 40). Thus, MCFs can be viewed as contributing to rapid leukemogenesis in situations where they can form.

OTHER STUDIES ON MuLV LEUKEMOGENESIS

Most of the molecular and virological experiments described above involved end-stage tumors. MuLV leukemogenesis has been studied from other perspectives as well.

Immunological Considerations

For viruses such as M-MuLV, which induce neoplasms of the immune system, some investigators have suggested immunological mechanisms in leukemogenesis. One model proposes that MCFs may induce leukemia by binding to MCF-specific T-lymphocytes and chronically stimulating them (30). One prediction of this model would be that M-MuLV-induced tumors should have functional T-cell receptors specific for MCF glycoprotein, but such findings have not been reported. Another study suggested that M-MuLV induces disease by inducing an anti-M-MuLV T-cell response, which results in chronic expansion of T-lymphopoiesis and development of leukemia (23). However, many M-MuLV-induced lymphomas are of immature thymocytes, which have not matured to where they respond to external antigen by activation and division.

Tumor Transplantation Experiment

Another approach to studying early events in leukemogenesis has been tumor transplantation. In these experiments, cells from different hematopoietic organs are transplanted from an infected preleukemic into a recipient, and the appearance of donor cell tumors is monitored (18). Studies of AKR or M-MuLV leukemias (both T-lymphoid) indicated that at early times after infection, hematopoietic organs such as bone marrow and spleen contain cells with leukemogenic potential, while the thymus does not (2, 18). In contrast, at later times, preleukemic cells could additionally be found in the thymus. Thus, early events in MuLV leukemogenesis may take place in bone marrow or spleen, even though leukemia ultimately develops in the thymus.

Physiological Changes during Leukemogenesis

During MuLV-induced T-lymphomagenesis, preleukemic physiological changes occur, most notably thymic atrophy and splenic enlargement; splenic enlargement may precede thymic atrophy somewhat. Haran-Ghera et al. (19) have attributed thymic atrophy to thymolytic effects of MCF recombinants in spontaneous AKR leukemia. Preleukemic splenomegaly is generally mild (two- to threefold enlargement for M-MuLV-inoculated mice at 6 to 8 weeks) and characterized by increased numbers of null cells (surface immunoglobulin, Thy 1 negative) (38). The null cells presumably represent hematopoietic

progenitors. There is also a positive correlation between virus-induced splenomegaly in different inbred mouse strains and their susceptibility to M-
MuLV leukemogenesis (38). Both the splenic enlargement and thymic atrophy suggest that virus-induced perturbations in hematopoiesis may be important for development of leukemia.

Mo+PyF101 M-MuLV AS A TOOL FOR STUDYING LEUKEMOGENESIS

Our investigations of M-MuLV leukemogenesis have been greatly aided
by an LTR variant of M-MuLV containing enhancer sequences from the
F101 strain of polyomavirus. The Mo+PyF101 LTR contains the polyomavirus F101 B enhancer element inserted into the wild-type M-MuLV LTR
at −150 bp, between the M-MuLV promoter-proximal sequences and enhancers (Fig. 2). This LTR was originally used to show that the M-MuLV
enhancers are not functional in undifferentiated embryonal carcinoma (EC)
cells (9, 29). In transient assays in F9 EC cells, the wild-type M-MuLV LTR
was inactive while the Mo+PyF101 LTR was active.

Infectious M-MuLV driven by the Mo+PyF101 LTR was generated by
molecular cloning and transfection. Since the Mo+PyF101 LTR has altered
activity, the pathogenic potential of the virus was of interest. Given the effects
of enhancers on MuLV pathogenicity, it seemed that Mo+PyF101 M-MuLV
might induce neoplasms of novel cell types. Surprisingly, when Mo+PyF101
M-MuLV was inoculated into neonatal NIH Swiss mice, its only effect was
greatly reduced leukemogenicity. In our initial experiments, leukemogenicity
was completely abolished (9). In subsequent experiments Mo+PyF101 M-
MuLV induced T-lymphoma in some but not all animals, but with a greatly
increased latent period (greater than 6 months, versus 2 to 3 months for wild-
type M-MuLV) (6).

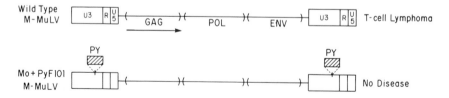

Figure 2. Wild-type and Mo+PyF101 M-MuLV. The genomes of the viruses are shown in DNA
forms. Wild-type M-MuLV is a standard replication-competent retrovirus that induces T-cell
lymphoma with a latency of 3 to 4 months under our conditions. Mo+PyF101 M-MuLV contains
enhancer sequences from the F101 mutant of polyomavirus inserted at −150 bp in the U3 region
of the M-MuLV LTR.

The greatly reduced leukemogenicity of Mo+PyF101 M-MuLV did not result from failure of the virus to establish infection in the animal (9). Therefore, in comparison to wild-type M-MuLV, Mo+PyF101 M-MuLV apparently has an altered repertoire of cell types for infection in the animal. In particular, some cell type has apparently been deleted from the repertoire which must be infected by wild-type M-MuLV in order for leukemia to develop. Thus, Mo+PyF101 M-MuLV is a useful tool for studying M-MuLV leukemogenesis.

One possible explanation for the nonleukemogenicity of Mo+PyF101 M-MuLV is that it does not efficiently infect thymocytes, the target cell for M-MuLV leukemia. However, when this was tested (11), Mo+PyF101 M-MuLV established comparable levels of infection in the thymus at preleukemic times (5 to 6 weeks) compared to wild-type M-MuLV-inoculated mice, with only a minimal (1 to 2 week) delay (Fig. 3). Thus, the nonleukemogen-

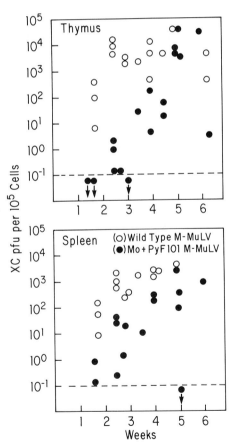

Figure 3. Thymic and splenic infection in inoculated mice. The levels of infection in thymus (A) and spleen (B) in NIH Swiss mice inoculated with wild-type and Mo+PyF101 M-MuLV are shown. Neonatal animals were inoculated subcutaneously with 2.5×10^4 XC PFU of either virus. Thymocytes and splenocytes were plated as infectious centers onto NIH 3T3 cells, and the amount of infectious virus was quantified by the UV-XC assay. All data points represent individual animals. Points with downward arrows indicate animals with undetectable virus levels (<0.1 XC PFU/10^5 cells). (Data taken from reference 11.)

icity of Mo+PyF101 M-MuLV could not be explained by an ability to infect the leukemic target cell.

PRELEUKEMIC HEMATOPOIETIC HYPERPLASIA IN THE SPLEEN

Since nonleukemogenic Mo+PyF101 M-MuLV could productively infect target thymocytes, it seemed that its defect might be in an earlier process in leukemogenesis. In light of the tumor transplantation experiments and increased splenic null cells, we tested preleukemic animals for increased hematopoiesis in the spleen, by agar colony assays for hematopoietic progenitors (10). As shown in Table 1, preleukemic NIH Swiss mice inoculated with wild-type M-MuLV showed 5- to 10-fold elevations in the spleen for myeloid and erythroid progenitors. Comparable agar colony assays for lymphoid progenitors are not currently available, but it seems likely that similar elevations occur for them as well. Thus, preleukemic mice show generalized hematopoietic hyperplasia in the spleen, with increased progenitors of multiple lineages.

Also as shown in Table 1, mice inoculated with Mo+PyF101 M-MuLV showed no splenic hyperplasia, even though splenocytes were extensively infected. This supports the importance of splenic hyperplasia to leukemogenesis, since the nonleukemogenic virus does not induce it. Moreover, the results suggest that Mo+PyF101 M-MuLV is defective in induction of splenic hyperplasia.

The major organ for hematopoiesis in adult mice is normally the bone marrow. Interestingly, when preleukemic mice were tested for elevated hematopoiesis in the bone marrow, little (one- to twofold) elevation in hematopoietic progenitors was observed (Table 1). This will be discussed below.

A TWO-INFECTION MODEL FOR M-MuLV LEUKEMOGENESIS

Based on these results, we proposed a two-infection model for M-MuLV leukemogenesis (10; Fig. 4). The first infection, occurring in bone marrow or spleen, results in generalized hematopoietic hyperplasia in spleen. The hyperplastic progenitors are potential target cells for leukemogenic transformation. In the case of M-MuLV, if a hyperplastic lymphoid progenitor migrates to the thymus and undergoes T-lymphoid differentiation, a second infection by M-MuLV (or an MCF recombinant) results in LTR activation of proto-oncogen, leading to T-lymphoma. Mo+PyF101 M-MuLV is incapable of carrying out the first infection leading to hematopoietic hyperplasia, but it can presumably carry out the second one, since it efficiently infects thymocytes in vivo.

Given the fact that the preleukemic hyperplasia involves multiple lineages, the model proposes that the LTR enhancer's influences on disease specificity occur during the second infection. For instance, a virus with enhancers that work efficiently in differentiated erythroid cells might select hyperplastic erythroid progenitors for LTR activation from the same pool of progenitors. As one test of this possibility, we studied animals inoculated with an LTR variant of M-MuLV, MF MuLV, in which the LTR from Friend erythroleukemia virus was substituted into M-MuLV. MF MuLV induces erythroleukemia, although all of the structural proteins come from M-MuLV (7). MF MuLV induced the same generalized hyperplasia that wild-type M-MuLV did; in particular, preleukemic FM MuLV-inoculated animals showed no higher elevations of erythroid progenitors than did animals infected with wild-type M-MuLV, even though the former virus induces erythroleukemia and the latter induces T-lymphoma (10). This was consistent with disease specificity being a function of the second infection.

STUDIES ON THE MECHANISM OF HYPERPLASIA INDUCTION

We have recently studied the mechanism by which M-MuLV induces splenic hyperplasia. Again, the Mo+PyF101 M-MuLV variant has been a useful tool.

Direct versus Indirect Effects on Progenitors

A fundamental question about hyperplasia induction is whether it results directly from infection of hematopoietic progenitors. For instance, an M-MuLV protein might be mitogenic in hematopoietic progenitors. On the other hand, M-MuLV might induce nonhematopoietic cells to produce hematopoietic growth factors such as interleukin-3. One way to distinguish between these possibilities would be to test whether hematopoietic stem cells in preleukemic animals are virus infected. If direct infection is required for hyperplasia, then all progenitors in preleukemic animals should be virus infected.

We tested hematopoietic progenitors (CFU_{mix}) from preleukemic animals inoculated with wild-type M-MuLV for virus infection by two methods: immunological staining for viral *gag* (p30 or CA) protein, or assaying for release of infectious M-MuLV in an XC plaque assay (5). As shown in Table 2, the great majority of CFU_{mix} from a preleukemic animal are virus infected, consistent with a direct infection mechanism for hyperplasia. However, studies on mice inoculated with Mo+PyF101 MuLV are very informative

Table 1. Hematopoietic Progenitors in Preleukemic Mice[a]

Expt no.	Virus inoculated	Weeks p.i.	Myeloid progenitors				Bone marrow: CFC/10^5 cells[e]
			Spleen				
			XC PFU/10^5 cells[b]	CFC/10^5 cells[c]	Total cells (10^8)[d]	Total CFC (10^3)	
1	None	6	ND[f]	0.6	2.7	1.6	13
	wt M-MuLV		ND	4.1 (6.8×)	2.3	9.4 (5.9×)	26 (2.8×)
2	None	7		2.1	2.1	4.4	67
	wt M-MuLV		ND	5.4 (2.6×)	2.8	15 (3.4×)	54 (0.8×)
3	None	7.5	ND	0.4	1.2	0.5	41
	wt M-MuLV			2.0 (5.0×)	3.0	6.0 (13×)	141 (3.4×)
4	None	8.5		1.6	2.6	4.2	
	wt M-MuLV		7.5 × 10^3	6.8 (4.3×)	2.6	18 (4.2×)	
	wt M-MuLV		ND	8.3 (5.2×)	4.3	36 (8.5×)	
	Mo+PyF101 M-MuLV		<10^0	1.2 (0.8×)	2.0	2.4 (0.6×)	
	Mo+PyF101 M-MuLV		4.5 × 10^2	2.3 (1.4×)	2.3	5.3 (1.3×)	
5	None	9		4.0	2.8	11.6	
	wt M-MuLV		1.8 × 10^4	11.5 (2.9×)	3.3	38 (3.4×)	
	Mo+PyF101 M-MuLV		3.2 × 10^3	2.5 (0.6×)	2.7	6.8 (0.6×)	
	Mo+PyF101 M-MuLV		<10^0	3.5 (0.9×)	3.0	10.5 (0.9×)	

Table 1. Continued

Expt no.	Virus inoculated	Weeks p.i.	Erythroid progenitors: spleen		
			CFU-E/10^5 cells[g]	Total cells (10^8)	Total CFU-E (10^3)
1	None	10.5	0.8	2.6	2.1
	wt M-MuLV		5 (6.3×)	4.7	23.5 (11×)
	wt M-MuLV		10.6 (13×)	3.6	38.2 (18×)
	Mo+PyF101 M-MuLV		0.8 (1×)	1.8	1.4 (0.7×)
	Mo+PyF101 M-MuLV		0.4 (0.5×)	2.2	0.9 (0.4×)

[a]Data from reference 10. Neonatal NIH Swiss mice were inoculated subcutaneously with 2.5×10^4 XC PFU of the indicated viruses (wt, wild type). Animals were sacrificed at the times indicated (weeks postinoculation [p.i.]). Single-cell suspensions were prepared from total spleen by grinding the organ through a fine wire mesh.

[b]Numbers of cells per 10^5 splenocytes which could act as infectious centers (by the XC plaque assay) when plated on NIH 3T3 cells.

[c]Splenocyte suspensions were assayed for myeloid colony-forming cells (CFC) as described in reference 10. Values shown indicate myeloid CFC per 10^5 cells. The concentration of CFC for the virus-inoculated animals relative to a control uninoculated animal in the same experiment is indicated in parentheses.

[d]Total number of cells recovered from the spleen was determined by counting a diluted sample with a hemacytometer.

[e]It was not possible to determine total numbers of bone marrow cells since recovery from the femurs was incomplete. Thus only the concentration of CFC is shown.

[f]ND, not done.

[g]Erythroid CFU (CFU-E) were assayed as described in reference 10. The criteria for scoring were relatively stringent; approximately 20 to 50 times more colonies showed at least one or two benzidine-positive cells and were potentially CFU-E as well. Thus the absolute numbers appear lower than reported by others.

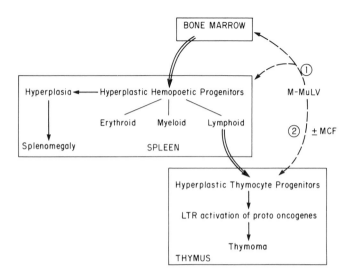

Figure 4. A two-infection model for M-MuLV leukemogenesis. In this model, M-MuLV infects the animal twice (dotted lines). The first infection occurs in the bone marrow or spleen and results in generalized hematopoietic hyperplasia and splenomegaly. If a hyperplastic lymphoid progenitor migrates to the thymus, it may become infected in a second event. This second infection would result in LTR activation of proto-oncogenes and outgrowth of the tumor. Double lines indicate the movement of hematopoietic stem cells between different organs. (From reference 10.)

(Table 2). CFU_{mix} from these animals also showed high levels of virus infection, even though the animals did not show hyperplasia. Thus, while infection of hematopoietic progenitors occurs at high frequency in M-MuLV-infected animals, infection per se is not sufficient to induce hyperplasia. Rather, some indirect mechanism must cause the hyperplasia.

Table 2. Assay for infectious virus in CFU_{mix} by cocultivation with NIH 3T3 cells[a]

Origin of CFU_{mix}	No. of CFU_{mix} assayed	No. of CFU_{mix} positive for infectious virus[b]
Uninfected animal	5	0
Wild-type M-MuLV-infected animal	17	15
Mo+PyF101 M-MuLV-infected animal	16	13
Wild-type M-MuLV assay plate, agar plug lacking CFU_{mix}	3[c]	0

[a]Individual 7-day CFU_{mix} colonies were picked and cocultivated with NIH 3T3 cells as described in the text. Agar plugs containing no CFU_{mix} were also picked to investigate ability of virus to spread through the agar. Data are from reference 5.
[b]Cultures were assayed for infectious ecotropic virus by the UV-XC plaque assay.
[c]Number of agar plugs assayed.

Studies in LTBMC

In an attempt to establish an in vitro system to study preleukemic hyperplasia, we established long-term bone marrow cultures (LTBMC) from normal and preleukemic mice (25). LTBMC support the growth of hematopoietic cells, particularly of the myeloid lineage (13). Surprisingly, LTBMC from preleukemic animals actually showed *less* hematopoiesis than those from control animals (Fig. 5). Microscopic examination indicated that this was due to a quantitative defect of the stromal cells. LTBMC are two-cell systems: stromal monolayers first establish, which then support proliferation of hematopoietic cells. The stromal cell defect in the LTBMC from preleukemic mice was quantitative rather than qualitative, since once a stromal monolayer was established (after a considerable lag), it was fully capable of supporting hematopoiesis.

LTBMC established from Mo+PyF101 M-MuLV-inoculated mice resembled cultures from uninoculated mice, and the stromal cell defect of LTBMC from wild-type M-MuLV-inoculated mice was not evident (25). This supported the relevance of the LTBMC defect to M-MuLV pathogenesis.

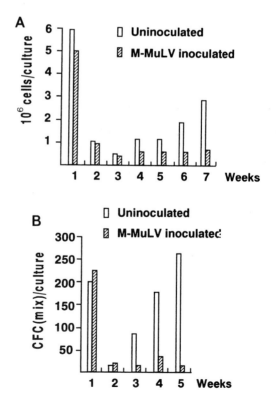

Figure 5. Hematopoietic cells in LTBMC from preleukemic mice inoculated with M-MuLV. (A) Total nonadherent cells in LTBMC from preleukemic mice. Mice 4 to 7 weeks old were sacrificed, and the bone marrow cells from each femur were collected and cultured in Corning T-25 flasks as described (25). One-half volume of medium was changed weekly, and nonadherent cells were counted with a hemacytometer. Each data point represents the average of results from two mice and four femurs. Open bars, LTBMC from uninoculated mice; hatched bars, LTBMC from preleukemic M-MuLV-inoculated mice. (B) Total CFC_{mix} in LTBMC from preleukemic mice. The nonadherent cells collected from the LTBMC of panel A were used in CFC_{mix} agar colony assays, as described previously (9). Averages of total CFC_{mix} per culture are shown. (Data taken from reference 25.)

How might the in vitro defect relate to M-MuLV pathogenesis in vivo? It should be noted that LTBMC are derived from bone marrow, not spleen. Moreover, our previous studies indicated that hematopoietic hyperplasia is pronounced in the spleen, but limited effects are evident in the bone marrow. One possible explanation could be that the LTBMC reflect an in vivo deficit of hematopoiesis in the bone marrow, which results in compensatory (extra-medullary) hematopoiesis in the spleen. Extramedullary hematopoiesis is a well-known physiological response to defects in bone marrow hematopoiesis.

The Role of MCF Recombinants

We also tested to see whether in vitro infection had effects on LTBMC. Infection of a biologically cloned M-MuLV stock (A9) onto LTBMC from an uninoculated animal led to reduced hematopoiesis (Fig. 6C), indicating that in vitro infection could lead to LTBMC stromal defects. During the course of these experiments, we became aware that the A9 stock of M-MuLV contained an MCF recombinant. Therefore, a parallel LTBMC culture was infected with an MCF-free stock of M-MuLV (43D) obtained by DNA trans-fection of a molecular clone of M-MuLV proviral DNA. Surprisingly, when the 43D stock was infected into the LTBMC, no inhibitory effects were seen (Fig. 6B). These results implicated MCF recombinants in the LTBMC defect.

The inhibitory effects of MCFs have been studied further (25). Similar, although less dramatic, effects were also evident upon infection of the A9 stock into NIH 3T3 fibroblasts. Further experiments indicated that combined infection of M-MuLV and an MCF recombinant was required, and the major effect was a decrease in cell growth rate (as opposed to cytolysis). The exact mechanism of the growth inhibition remains to be determined.

Our current working hypothesis for the generation of preleukemic hy-perplasia is as follows. Early during infection, M-MuLV generates an MCF recombinant. Combined infection of M-MuLV and MCF in the bone marrow stroma results in decreased bone marrow hematopoiesis, which leads to ex-tramedullary hematopoiesis and hyperplasia in the spleen. The splenic hy-perplasia could yield increased numbers of normal progenitor cell targets for later steps in leukemogenesis (e.g., LTR activation of proto-oncogenes). Al-ternatively, hyperplastic hematopoietic cells in the spleen might qualitatively differ from bone marrow progenitors, with increased potential for leuke-mogenic transformation. The tumor transplantation experiments identifying

---

Figure 6. LTBMC infected by different M-MuLV stocks. LTBMC were carried for 3 weeks. (A) Uninfected LTBMC from an uninoculated 7-week-old mouse. (B) A parallel LTBMC infected after establishment by the molecularly cloned M-MuLV (43D) stock. (C) LTBMC infected by the M-MuLV (A9) stock. Bar, 10 μm. (Data taken from reference 25.)

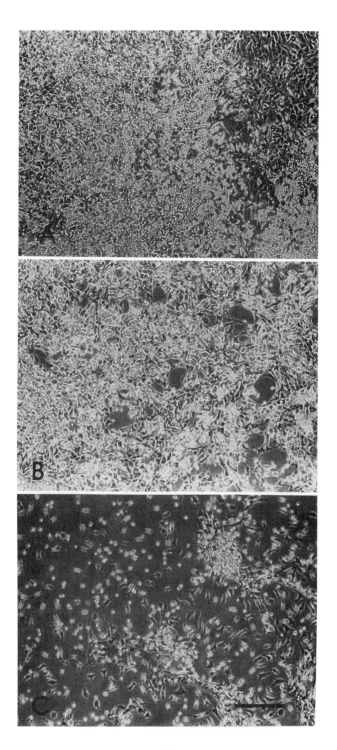

169

preleukemic cells in the spleen and marrow might support the latter possibility.

It should be noted that while this model proposes a role for MCF recombinants early in the leukemogenic process, MCFs may participate in later events as well. Indeed, MCF proviruses have been shown to activate proto-oncogenes (35).

The apparent virus-induced bone marrow defect in preleukemic animals might be analogous to other non-virus-induced preleukemic states. In humans, dysfunction of bone marrow hematopoiesis has been associated with preleukemic conditions (28). In the mouse, ablation of bone marrow hematopoiesis with the bone-seeking isotope ^{89}Sr results in marked splenic hyperplasia and increases in CFU_{mix} (1). Moreover, ^{89}Sr can induce leukemia in mice (44).

Basis for the Nonleukemogenicity of Mo+PyF101 M-MuLV

In the context of these results, it was interesting to determine why Mo+PyF101 M-MuLV does not induce preleukemic hyperplasia. Recently, we found that Mo+PyF101 M-MuLV does not generate detectable MCF recombinants when injected into animals, in contrast to wild-type M-MuLV (6). This is completely consistent with the role of MCF recombinants in inducing preleukemic hyperplasia and with the inability of Mo+PyF101 M-MuLV to do so.

The inability of Mo+PyF101 M-MuLV to form recombinants in vivo has been investigated further (6). Perhaps Mo+PyF101 M-MuLV cannot form recombinants in animals due to the inability to infect cells that transcribe endogenous polytropic (MCF-type) sequences. Infection of such cells is probably necessary for formation of heterozygous virions containing ecotropic and endogenous polytropic RNAs, which can yield an MCF recombinant genome during the next cycle of reverse transcription. Alternatively, MCF recombinants of Mo+PyF101 M-MuLV might form at the usual frequency, but not propagate efficiently. To test this, an MCF recombinant of Mo+PyF101 M-MuLV was generated by molecular cloning and transfection. This virus grew well in culture; however, when injected into newborn mice (either alone or in a mixture with Mo+PyF101 M-MuLV), it did not propagate, indicating the second mechanism.

ESTABLISHMENT OF THE PRELEUKEMIC STATE BY NONVIRAL MEANS

As described in our model of M-MuLV leukemogenesis (Fig. 4), Mo+PyF101 M-MuLV is defective in generation of preleukemic hyperplasia.

Moreover, our model suggests that Mo+PyF101 M-MuLV should be capable of carrying out the second infection and could induce leukemia if preleukemic hyperplasia is induced by other means. To test this, we studied animals inoculated with ^{89}Sr (26). As mentioned above, administration of ^{89}Sr results in ablation of bone marrow hematopoiesis and extramedullary hematopoiesis in the spleen. Neonatal NIH Swiss mice were inoculated with Mo+PyF101 M-MuLV, followed by injection with four weekly doses of ^{89}Sr beginning at 6 weeks. This resulted in efficient and rapid induction of T-lymphoma; greater than 80% of the animals developed disease, with a mean time of 17 weeks. Under similar conditions, inoculation with wild-type M-MuLV resulted in leukemia with a mean time of 12 to 13 weeks. Either ^{89}Sr or Mo+PyF101 M-MuLV alone yielded disease in fewer animals and with markedly longer time courses. Further analysis of the resulting tumors confirmed that they were infected with Mo+PyF101 M-MuLV. Moreover, in a significant number of cases ($\sim$33%), Mo+PyF101 M-MuLV proviruses were found inserted at common integration sites for wild-type M-MuLV tumors, including c-*myc*, *pvt-1*, and *pim-1*. Thus, induction of extramedullary hematopoiesis by non-viral means can complement the leukemogenic defect of Mo+PyF101 M-MuLV. Moreover, the results confirmed that Mo+PyF101 M-MuLV is competent in carrying out late steps in leukemogenesis such as LTR activation of proto-oncogenes.

SUMMARY

The work reviewed here highlights several aspects of leukemogenesis by M-MuLV, particularly preleukemic events, as follows.

(i) M-MuLV leukemogenesis is a multistep process in which the virus infects the host at least twice: first inducing preleukemic hematopoietic hyperplasia in the spleen and later causing LTR activation of proto-oncogenes in the tumor cells.

(ii) MCF recombinants play an early role in the leukemogenic process. Combined infection of MCF and ecotropic M-MuLV causes growth inhibitory effects for bone marrow stromal cells as measured in vitro in LTBMC. Defects in bone marrow hematopoiesis in vivo caused by combined infection may result in the observed splenic hyperplasia (extramedullary hematopoiesis).

(iii) The nonleukemogenic Mo+PyF101 M-MuLV has been instrumental in elucidating preleukemic events induced by M-MuLV. Mo+PyF101 M-MuLV does not induce preleukemic hyperplasia in the spleen, apparently due to an inability to propagate MCF recombinants in the animal. However, if splenic hyperplasia is induced by other means, Mo+PyF101 M-MuLV is capable of carrying out later steps in leukemogenesis.

Acknowledgments. This work was supported by National Institutes of Health grant CA32455 and by the University of California-Irvine Cancer Research Institute. B.K.B. was supported by a postdoctoral fellowship from the National Cancer Center, and B.R.D. was supported by a postdoctoral fellowship from the American Cancer Society, California Division.

REFERENCES

1. **Adler, S., F. Trobaugh, Jr., and W. Knospe.** 1977. Hemopoietic stem cell dynamics in ^{89}Sr marrow-ablated mice. *J. Lab. Clin. Med.* **89:**592–602.
2. **Asjo, B., L. Skoog, I. Palminger, F. Wiener, D. Asaak, J. Cerny, and E. M. Fenyo.** 1985. Influence of genotype and the organ of origin on the subtype of T-cell in Moloney lymphomas induced by transfer of preleukemic cells from athymic and thymus-bearing mice. *Cancer Res.* **45:**1040–1045.
3. **Bosselman, R. A., F. van Straaten, C. Van Beveren, I. M. Verma, and M. Vogt.** 1982. Analysis of the *env* gene of a molecularly cloned and biologically active Moloney mink cell focus-forming proviral DNA. *J. Virol.* **44:**19–31.
4. **Breuer, M. L., H.T. Cuypers, and A. Berns.** 1989. Evidence for involvement of *pim-2*, a new common proviral insertion site, in progression of lymphomas. *EMBO J.* **8:**743–748.
5. **Brightman, B. K., B. R. Davis, and H. Fan.** 1990. Preleukemic hematopoietic hyperplasia induced by Moloney murine leukemia virus is an indirect consequence of viral infection. *J. Virol.* **64:**4582–4584.
6. **Brightman, B. K., A. Rein, D. Trepp, and H. Fan.** *Proc. Natl. Acad. Sci. USA*, in press.
7. **Chatis, P. A., C. A. Holland, J. W. Hartley, W. P. Rowe, and N. Hopkins.** 1983. Role for the 3′ end of the genome in determining disease specificity of Friend and Moloney murine leukemia virus. *Proc. Natl. Acad. Sci. USA* **80:**4408–4411.
8. **Cuypers, H. T., G. Selten, W. Quint, M. Ziljstra, E. R. Maandag, W. Boelens, P. van Wezenbeek, C. Melieft, and A. Berns.** 1984. Murine leukemia virus T-cell lymphomagenesis: integration of proviruses in a distinct chromosomal region. *Cell* **37:**141–150.
9. **Davis, B., E. Linney, and H. Fan.** 1985. Suppression of leukemia virus pathogenicity by polyoma virus enhancers. *Nature* (London) **314:**550–553.
10. **Davis, B. R., B. K. Brightman, K. G. Chandy, and H. Fan.** 1987. Characterization of a preleukemic state induced by Moloney murine leukemia virus: evidence for two infection events during leukemogenesis. *Proc. Natl. Acad. Sci. USA* **84:**4875–4879.
11. **Davis, B. R., K. G. Chandy, B. K. Brightman, S. Gupta, and H. Fan.** 1986. Effects of non-leukemogenic and wild-type Moloney murine leukemia virus on lymphoid cells in vivo: identification of a preleukemic shift in thymocyte subpopulations. *J. Virol.* **60:**204–214.
12. **desGrossiellers, L., E. Rassart, and P. Jolicoeur.** 1983. Thymotropism of murine leukemia virus is conferred by its long terminal repeat. *Proc. Natl. Acad. Sci. USA* **80:**4203–4207.
13. **Dexter, T. M., T. D Allen, and L. G. Lajtha.** 1977. Conditions controlling the proliferation of hemopoietic stem cells in vitro. *J. Cell. Physiol.* **91:**335–344.
14. **Evans, L. H., and F. G. Malik.** 1987. Class II polytropic murine leukemia viruses (MuLVs) of AKR/J mice: possible role in the generation of class I oncogenic polytropic MuLVs. *J. Virol.* **61:**1882–1892.
15. **Fan, H., H. Chute, E. Chao, and P. K. Pattengale.** 1988. Leukemogenicity of Moloney murine leukemia viruses carrying polyoma enhancer sequences in the long terminal repeat is dependent in the nature of the inserted polyoma sequences. *Virology* **166:**58–65.

16. **Graham, M., J. M. Adams, and S. Corey.** 1985. Murine T lymphomas with retroviral inserts in the chromosome 14 locus for plasmacytoma variant translocations. *Nature* (London) **314:**740–745.

17. **Hanecak, R., P. K. Pattengale, and H. Fan.** 1988. Addition or substitution of simian virus 40 enhancer sequences into the Moloney murine leukemia virus (M-MuLV) long terminal repeat yields infectious M-MuLV with altered biological properties. *J. Virol.* **62:**2427–2436.

18. **Haran-Ghera, N.** 1980. Potential leukemic cells among bone marrow cells of young AKR/J mice. *Proc. Natl. Acad. Sci. USA* **77:**2923–2926.

19. **Haran-Ghera, N., A. Peled, F. Leef, A. D. Hoffman, and J. A. Levy.** 1987. Enhanced AKR leukemogenesis by the dual tropic viruses. I. The time and site of origin of potential leukemic cells. *Leukemia* **1:**442–449.

20. **Hartley, J. W., N. K. Wolford, L. J. Old, and W. P. Rowe.** 1977. A new class of murine leukemia virus associated with development of spontaneous lymphomas. *Proc. Natl. Acad. Sci. USA* **74:**789–792.

21. **Hayward, W. S., B. G. Neel, and S. M. Astrin.** 1981. Activation of a cellular *onc* gene by promoter insertion in ALV-induced lymphoid leukosis. *Nature* (London) **290:**475–480.

22. **Lazo, P. A., J. S. Lee, and P. N. Tsichlis.** 1990. Long-distance activation of the *Myc* proto-oncogene by provirus insertion at *Mlvi-1*, or *Mlvi-4* in rat T-cell lymphomas. *Proc. Natl. Acad. Sci. USA* **87:**170–173.

23. **Lee, J. C., and J. N. Ihle.** 1981. Chronic immune stimulation is required for Moloney leukemia virus-induced lymphomas. *Nature* (London) **289:**407–409.

24. **Lenz, J., D. Celander, R. L. Crowther, R. Patarca, D. W. Perkins, and W. A. Haseltine.** 1984. Determination of the leukemogenicity of a murine retrovirus by sequences with the long terminal repeat. *Nature* (London) **308:**467–470.

25. **Li, Q., and H. Fan.** 1990. Combined infection by Moloney murine leukemia virus and a mink cell focus-forming virus recombinant induces cytopathic effects in fibroblasts or in long-term bone marrow cultures from preleukemic mice. *J. Virol.* **64:**3701–3711.

26. **Li, Q.-X., and H. Fan.** Submitted for publication.

27. **Li, Y., C. A. Holland, J. W. Hartley, and N. Hopkins.** 1984. Viral integration near *c-myc* in 10 to 20% of MCF 247-induced AKIR lymphomas. *Proc. Natl. Acad. Sci. USA* **81:**6808–6811.

28. **Lichtman, M. A., and J. K. Brennan.** 1983. Dyshemopoeitic (preleukemic) disorders, p. 175–184. *In* W. J. Williams, E. Beutler, A. J. Erslev, and M. A. Lichtman (ed.), *Hematology*, 3rd ed. McGraw-Hill Book Co., New York.

29. **Linney, E., B. Davis, J. Overhauser, E. Chao, and H. Fan.** 1984. Non-function of a Moloney murine leukaemia virus regulatory sequence in F9 embryonic carcinoma cells. *Nature* (London) **308:**470–472.

30. **McGrath, M. S., and I. L. Weissman.** 1979. AKR leukemogenesis: identification and biological significance of thymic lymphoma receptors for AKR retroviruses. *Cell* **17:**65–75.

31. **Mucenski, M. L., D. J. Gilbert, B. A. Taylor, N. A. Jenkins, and N. G. Copeland.** 1987. Common sites of viral integration in lymphomas arising in AKXC recombinant inbred mouse strains. *Oncogene Res.* **2:**33–48.

32. **Nusse, R., and H. E. Varmus.** 1982. Many tumors induced by the mouse mammary tumor virus contain a provirus integrated in the same region of the host genome. *Cell* **31:**99–109.

33. **Payne, G. S., J. M. Bishop, and H. E. Varmus.** 1982. Multiple arrangements of viral DNA and an activated host oncogene in bursal lymphomas. *Nature* (London) **295:**209–214.

34. **Ruscetti, S., R. Matthai, and M. Potter.** 1985. Susceptibility of Balb/c mice carrying various DBA/2 genes to development of Friend murine leukemia virus-induced erythroleukemia. *J. Exp. Med.* **162:**1579–1587.

35. **Selten, B., H. T. Cuypers, M. Ziljstra, C. Blief, and A. Berns.** 1984. Involvement of *c-myc* in M-MuLV-induced T-cell lymphomas of mice: frequency of activation. *EMBO J.* **13:**3215–3222.

36. **Short, M. K., S. A. Okenquist, and J. Lenz.** 1987. Correlation of leukemogenic potential of murine retroviruses with transcriptional tissue preference of the viral long terminal repeats. *J. Virol.* **61:**1067–1072.

37. **Steffen, D.** 1984. Proviruses are adjacent to *c-myc* in murine leukemia virus-induced lymphomas. *Proc. Natl. Acad. Sci. USA* **81:**2097–2101.

38. **Storch, T. G., P. Arnstein, U. Manohar, W. M. Leiserson, and T. M. Chused.** 1985. Proliferation of infected lymphoid precursors before Moloney murine leukemia virus-induced T-cell lymphoma. *J. Natl. Cancer Inst.* **74:**137–143.

39. **Thomas, C. Y., and J. M. Coffin.** 1982. Genetic alterations of RNA leukemia viruses associated with the development of spontaneous thymic leukemia in AKR/J mice. *J. Virol.* **43:**416–426.

40. **Tsichlis, P. N.** 1987. Oncogenesis by Moloney murine leukemia virus. *Anticancer Res.* **7:**171–180.

41. **Tsichlis, P. N., M. A. Lohse, C. Szpirer, J. Szpirer, and G. Levan.** 1985. Cellular DNA regions involved in the induction of rat thymic lymphomas (*Mlvi-1*, *Mlvi-2*, *Mlvi-3*, and *c-myc*) represent independent loci as determined by their chromosomal map location in the rat. *J. Virol.* **56:**938–942.

42. **vander Putten, H., W. Quint, J. van Raaji, E. R. Maandag, I. M. Verma, and A. Berns.** 1981. M-MuLV-induced leukemogenesis: integration and structure of recombinant proviruses in tumors. *Cell* **24:**729–739.

43. **Vijaya, S., D. L. Steffen, C. Kozak, and H. L. Robinson.** 1987. *Dsi-1*, a region with frequent proviral insertions in Moloney murine leukemia virus-induced rat thymomas. *J. Virol.* **61:**1164–1170.

44. **Wright, J. F., M. Goldman, and L. K. Bustad.** 1972. Age, strain, and dose-rate effects of ^{89}Sr in mice, p. 168–181. *In* M. Goldman and L. Bustad (ed.), *Biomedical Implications of Radiostrontium Exposure.* U.S. Atomic Energy Commission Office of Information Service, Washington, D.C.

45. **Yoshimura, F. K., B. Davison, and K. Chaffin.** 1985. Murine leukemia virus long terminal repeat sequences can enhance gene activity in a cell-type-specific manner. *Mol. Cell. Biol.* **5:**2832–2835.

Viruses That Affect the Immune System
Edited by Hung Y. Fan et al.
© 1991 American Society for Microbiology, Washington, DC 20005

Chapter 12

Endogenous Murine Retroviruses and Leukemia

John M. Coffin, Jonathan P. Stoye, and Wayne N. Frankel

One of the most complex systems in virology was brought to light by the development of strains of inbred mice, such as AKR, with a high incidence of spontaneous thymic lymphoma (10). It was from such mice that the first murine leukemia virus (MLV) was isolated (13) and subsequently shown to be the somewhat modified product of an endogenous provirus inducible from uninfected cells of AKR and other mice (14). A further analysis of the virus associated with the AKR disease (known as mink cell focus-forming [MCF] virus) revealed that it had undergone characteristic genetic alterations. The changes include the replacement of sequence in *env*, leading to a virus of altered host range. The initial virus is an ecotropic virus which utilizes a receptor largely limited to mouse cells, whereas the recombinant virus is polytropic, utilizing a receptor found on cells of many mammals and birds (30, 31). Also included are changes in the U3 portion of the long terminal repeat (LTR) (12, 21) which can confer increased expression ability in the thymic target cells (17). Both these changes are the consequence of recombination events between the replicating ecotropic virus and at least two nonecotropic proviruses, one of which donates the host range determinant (28) and the other the new LTR sequence (12). At the time this project was initiated, the endogenous, nonecotropic provirus donors of these other se-

---

John M. Coffin and Wayne N. Frankel • Department of Molecular Biology and Microbiology, Tufts University School of Medicine, 136 Harrison Avenue, Boston, Massachusetts 02111. **Jonathan P. Stoye** • National Institute for Medical Research, The Ridgeway Mill Hill, London NW7 1AA, England.

quences had not been identified, but were known to be members of a large family of related proviruses in mice. In the work described here, we present a detailed analysis of these proviruses, their distribution in inbred mice, their chromosomal locations, their value as tools for genetic analysis, and their effects on the host, both as mutagens and as participants in the complex evolution process which creates oncogenic viruses from benign ancestors.

TYPES OF NONECOTROPIC PROVIRUSES

When examined with relatively nonspecific MLV DNA probes, Southern blots of mouse DNA reveal a bewildering complexity of bands, indicating many more proviruses than can be readily distinguished or enumerated (1, 24). To resolve this problem, we set out initially to clone and analyze as many proviruses as possible from one strain of inbred mice (HRS/J). Detailed molecular analysis (including nucleotide sequencing of *env* and LTR regions) revealed that the complex mixture could be resolved into three classes of provirus in addition to the ecotropic virus, each containing 10 to 20 very closely related members (less than 1% sequence divergence, excluding simple deletions). The three groups differed in several features (Fig. 1), including some diagnostic restriction enzyme cleavage sites, a specific 110-base insertion into U3, and a specific sequence variation in the *env* gene. The *env* gene sequences allowed us to classify two of the types as encoding Env proteins (previously identified from recombinant or induced viruses) as xenotropic and polytropic (28). The third type had not previously been recognized (although it does appear in some recombinants derived from Friend MLV [5, 20, 28]), and we named it modified polytropic, since it is quite similar to the

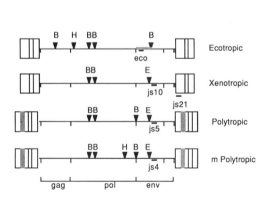

Figure 1. The four classes of endogenous MLV proviruses. The figure indicates the structure of the ecotropic as well as the three classes of nonecotropic endogenous proviruses defined by this work (m Polytropic, modified polytropic). Single letters denote restriction enzyme cleavage sites (B, *Bam*HI; H, *Hind*III; E, *Eco*RI). The standard band shown at bottom represents an insert sequence found in polytropic modified polytropic proviruses. The bases underneath *env* and LTR regions show reactivity with specific oligonucleotide probes.

polytropic protein and, when inserted into a replicating virus, encodes a similar (but not identical) host range (data not shown).

Although the sequences of these three provirus types were quite similar, we were able to identify a small (27-base) nucleotide sequence polymorphism in the SU (gp70) region of *env* with sufficient divergence to allow preparation of oligonucleotide probes specific for each type. Together, these probes detect virtually all of the nonecotropic proviruses seen by less specific probes (9, 29) (data not shown); separately, they each react with a sufficiently small set of proviruses that all members can be distinguished in a typical Southern blot. The strategy for their use is shown in Fig. 2. Note that the site of the probe is immediately 3' of *Eco*RI and *Pvu*II sites shared by all three groups and that no other sites for these two enzymes lie between the probe site and the end of the provirus. Note also that the size of the viral fragment is fairly short (less than 2 kb), ensuring a good spread of the bands whose size depends

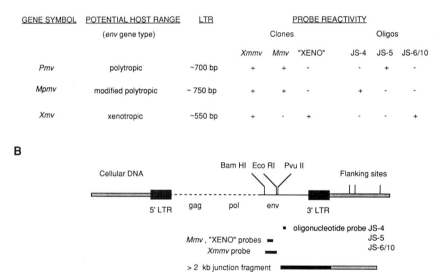

Figure 2. (A) Structural characteristics of *Pmv, Mpmv,* and *Xmv* loci. *Xmmv* refers to 5' *env* region probes utilized previously by other groups. *Mmv* and "XENO" refer to MCF-"specific" or *Xmv* probes, respectively. (B) Prototype restriction map based on modified polytropic proviruses cloned from HRS/J mice, depicting proviral-cellular DNA junction fragments created by restriction enzymes *Eco*RI, *Pvu*II, or *Bam*HI. LTRs are shown as black boxes, the viral genome is shown as a solid or a dashed line, and cellular DNA is shown as a stippled bar. Several other *Bam*HI and *Pvu*II restriction sites, but no *Eco*RI sites, are present in the proviral region, depicted as a dashed line. The locations of respective *env* probes are also shown.

primarily on the distance to the next *Eco*RI or *Pvu*II site from the point of integration.

The use of these probes to detect specific endogenous proviruses is illustrated in Fig. 3 for a set of common inbred strains. It is readily apparent that every strain has a unique and characteristic pattern of endogenous proviruses (which we have found to provide a useful "fingerprint" in cases of mixed strain identity) and that very few proviruses are shared by all strains. Indeed, in a much larger survey of inbred strains (not shown) we have found only one provirus to be shared among all. This diversity clearly reflects the recent insertion of these elements into the mouse germ line and the diversity in the original breeding stock which gave rise to the modern inbred lines.

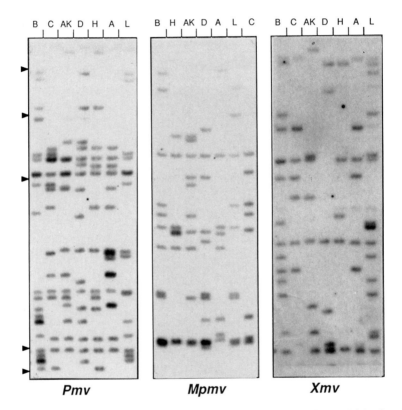

Figure 3. A comparison of the *Pmv, Mpmv,* or *Xmv* proviral content of several inbred strains of mice. Shown is a Southern blot of *Pvu*II-digested genomic DNA from inbred strains hybridized to oligonucleotide probe. Inbred strain abbreviations used: A, A/J; AK, AKR/J; B, C57BL/6J; C, BALB/cJ; D, DBA/2J; H, C3H/HeJ; L, C57L/J. Arrowheads show positions of *Hin*dIII-digested λ DNA molecular size standards: 9.4, 6.6, 4.3, 2.3, and 2.0 kb, in descending order.

GENETIC MAPPING OF PROVIRUSES

The polymorphism of integration sites of endogenous nonecotropic proviruses provides the means for determining their chromosomal location via standard genetic techniques. This was accomplished using mice from seven sets of recombinant inbred (RI) lines, each created by inbreeding multiple lines from the progeny of an original mating between two inbred lines (7–9). From the strain distribution pattern of each proviral junction fragment within such a set (for example, see Fig. 4), the corresponding provirus could be linked to previously mapped anchor loci whose strain distribution pattern within that set had previously been determined. In this way, we generated the map shown in Fig. 5 (9). This map gives the inferred chromosomal location of about 116 proviruses in inbred mouse strains. These proviruses

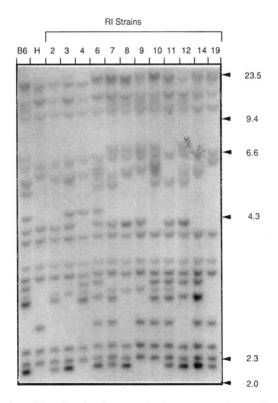

Figure 4. Segregation of *Pmv* junction fragments in the BXH set of RI strains. *Eco*RI-digested genomic DNA from BXH RI strains plus progenitor strains C57BL/6J and C3H/HeJ was separated by electrophoresis and hybridized to JS5. *Hin*dIII-digested λ DNA molecular size standards are shown.

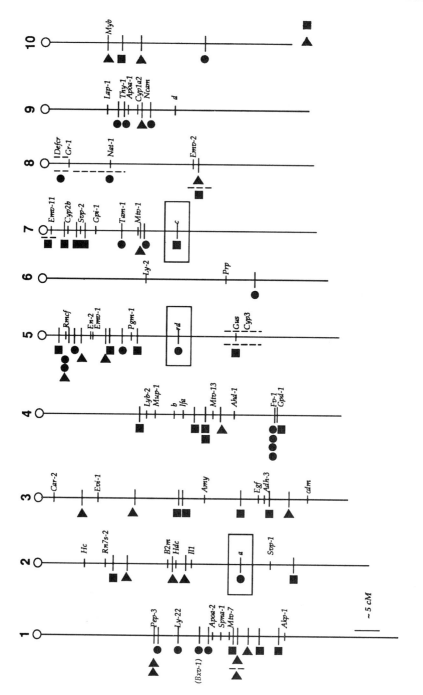

~ 5 cM

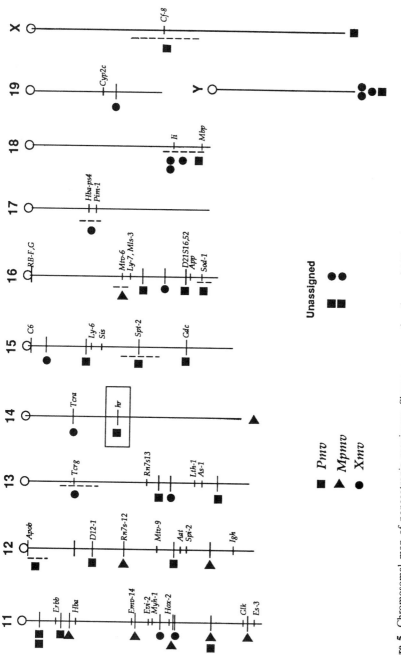

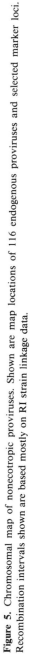

Figure 5. Chromosomal map of nonecotropic proviruses. Shown are map locations of 116 endogenous proviruses and selected marker loci. Recombination intervals shown are based mostly on RI strain linkage data.

now constitute about 6% of mapped mouse genetic markers. Although most of the relative map locations were determined by using the nonstandard recombinant inbred approach, we have confirmed many of them with additional evidence from more traditional backcross and intercross analyses (6).

Several points of interest became apparent during the course of generating this map.

First, with one interesting exception, the proviruses are apparently randomly distributed throughout the genome, at least one per chromosome, and with no evidence of clustering or other nonrandom patterns.

Second, virtually all the polymorphism observed represented distinct integrations. We found no evidence for variation due to fragment length polymorphism of an individual provirus (i.e., apparently different proviruses mapping to exactly the same location).

Third, the proviruses are quite stable. In our survey, only a few instances of fragments not found in either parent or of loss of fragments common to both parents were seen (Table 1). From this we can infer a rate of around 4 $\times$ 10^{-6} per meiotic generation for proviral loss and a rate of gain of 1 provirus per 3,500 generations. We presume that the latter reflects additional germ line infections, as has been shown in other cases (2, 27), but have not tested this point directly. The rate of loss of proviruses is quite similar to that reported earlier (25) for a single ecotropic provirus in a large number of mice. While there are many ways in which reactive bands could disappear and we have not examined the mechanism of this loss, in other cases it has been shown to be primarily due to recombination between the LTRs (2, 27).

Fourth, the proviruses we have delineated provide an extremely valuable set of markers for mouse geneticists. In the correct cross (e.g., an intercross between C57BL/6J and DBA/2J), we can survey well over 80% of the genome with as few as 25 progeny mice, using two restriction enzyme digestions and three probes (Fig. 6). As an example of this, we have been able to localize an immunoglobulin H transgene which undergoes class switching with the resistant immunoglobulin loci by its close linkage to two proviruses in chromosome 5, thus demonstrating transchromosomal isotype switching (11).

Thus, analysis of these proviruses has not only been highly informative regarding the biology of their association with mice, but has provided a useful set of distributed markers for genetic analysis. The interested reader can obtain more information and detailed protocols directly from us.

ASSOCIATION WITH MUTATIONS

The insertion of an endogenous provirus into the germ line of mice is an obvious mutagenic event. Indeed, it has been shown that an ecotropic

Table 1. Stability and Acquisition of Noncecotropic MLVs in RI Strains

RI set	No. of filial generations	No. of proviruses surveyed[a]	Proviruses lost[b]			Proviruses gained[c]		
			Xmv	Mpmv	Pmv	Xmv	Mpmv	Pmv
AXB,BXA	>1,031	21	#9 (AXB-7)					
AKXD	1,369	19			#20 (AKXD-14)	#1 (AKXD-2)		
AKXL	1,277	13						
BXD	1,963	20			#6 (BXD-2,-30)		#1 (BXD-25)	
BXH	934	17						
CXB	488	23						

[a] ~27% of all proviruses.
[b] Average rate of loss, ~1/250,000 per provirus per generation.
[c] Average rate of gain, ~1 provirus per 3,500 generations.

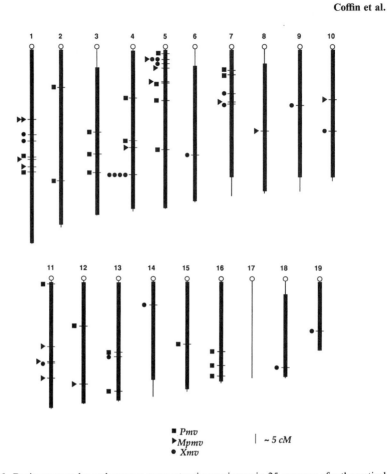

Figure 6. Regions swept by endogenous nonecotropic proviruses in 25 progeny of a theoretical C57BL/6J × DBA/2J intercross. Thick lines represent chromosomal regions swept by proviral loci.

provirus insertion caused the *d* (dilute brown) mutation in mice (3) and that insertion of Moloney MLV proviruses introduced deliberately can also be mutagenic (15). To test the extent to which this much larger set of proviruses has contributed to genetic variability in the mouse, we assessed the linkage of specific mutants to specific proviruses by analysis of the RI data as well as a number of congenic and mutant inbred lines. We found a suggestive linkage for about 5% of mutations screened (data not shown). (A long list of mutations unassociated with proviruses is available on request.) Of the mutations shown, several (*c, a, rd*) are well known and extensively studied (22), and we are collaborating with others on a closer analysis of them. Indeed,

Xmv-28 is inserted in an intron of the *rd* gene (unpublished results). Of particular virological interest is the *Fv-1* locus, which restricts the replication of "mismatched" MLVs due to a poorly understood intracellular effect (19, 32). There is a striking association between the *Fv-1*[b] allele and a cluster of four *Xmv* proviruses on chromosome 4. This association raised the possibility that the *Fv-1* effect might be mediated via a product of one or more of these proviruses acting, for example, to compete with a cellular factor necessary for virus replication. Whether or not this proviral linkage to *Fv-1* is merely a chance phenomenon still needs to be resolved.

The foremost association of insertion of endogenous proviruses with mutation was found in the case of the *hr* mutation carried by HRS/J and some other strains of inbred mice. Homozygous *hr/hr* mice have a normal initial hair growth shortly after birth, but lose their hair permanently at about 2 weeks of age (4). Unlike the *nu* mutation, *hr* is not associated with a profound immunological defect, although it is linked to a high incidence of thymoma, virologically similar to that of AKR (12). In initial experiments, we observed that a specific polytropic provirus, *Pmv-43*, was linked to *hr* in all strains tested. Genetic proof that the proviruses were causally linked to this mutation came from analysis of a revertant line isolated in Australia (29). Analysis of the proviruses of these (now hairy) mice, using a probe for cell sequence flanking *Pmv-43*, revealed that the revertant did not have an intact provirus, but had retained a solo LTR which was identical in sequence to that of the provirus. The apparently simultaneous loss of the provirus (by recombination across the LTRs) and reversion of the mutation are both extremely rare events and thus provide compelling evidence for causality. This also implies that the provirus is in a noncoding region, perhaps an intron, of the *hr* gene. The insertion site for *Pmv-43* thus serves as a starting point to identify and clone *hr*. By "walking" from the insertion site through a genomic library using a series of flanking sequence probes, we have identified several DNA clones with the expected characteristics (Fig. 7): a CG-rich region like that of many eukaryotic promoter regions, a region conserved in many mammalian species, and two candidate exons for which probes detect transcripts in skin (and, to a less extent, brain) of young mice, but not in other tissues.

GENERATION OF LEUKEMOGENIC VIRUSES

A major initial goal of this project was to identify the nonecotropic proviruses involved in the generation of leukemogenic viruses. To date, in fact, we have been unable to identify which of the 17 *Pmv* proviruses in AKR is indeed the donor of the polytropic *env* gene. The problem is twofold. First,

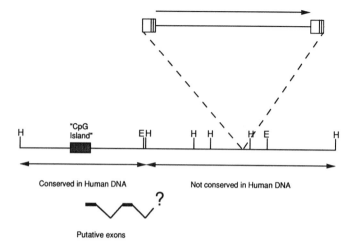

Figure 7. The *hr* locus. The map summarizes the current status of our knowledge concerning the structure of the *hr* locus, including some restriction enzyme sites and the insertion site of the *Pmv-43* provirus.

the sequences of all *Pmv env* genes are so similar to one another and to that of recombinant (MCF) viruses that almost any one could be a donor. Second, while we can exclude many *Pmv* loci, we have been unable to make a genetic association of a specific provirus with the ability to generate MCF viruses even in recombinant inbred sets where this characteristic is segregating (8).

We were more fortunate with the identification of the LTR donor to the recombinant viruses. A specific oligonucleotide derived from predicted U3 sequence (called JS21) identifies sequences that are found in only one provirus in normal tissue, but which are greatly amplified in thymomas (Fig. 8). Moreover, these amplified proviruses yield the same size restriction fragments as newly acquired fragments hybridizing to the *Pmv* probe. We thus conclude that only one endogenous provirus LTR sequence becomes part of the MCF LTR. Analysis of the distribution of this LTR in inbred and congenic strains of mice has revealed that it is identical to the provirus *Bxv-1*, which can be induced to give replication-competent xenotropic MLV (14). Thus, the new proviruses which are found integrated in the lymphomas and are apparently the proximal cause of disease (presumably by insertion adjacent to one or more proto-oncogenes [23]) are derived from three parents—an ecotropic, polytropic, and xenotropic provirus—and at least two recombinant events, each requiring a double crossover (21).

The two crossovers are not the only events which conspire to create the leukemogenic virus. Like many MLVs, MCF viruses are usually characterized

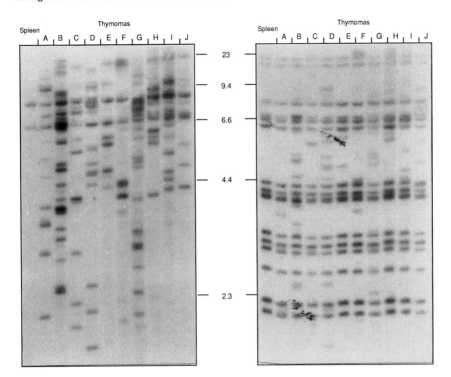

Figure 8. Somatically acquired proviruses in AKR tumors. Each panel contains *Eco*RI-digested DNA from control tissue (spleen) as well as 10 independent thymomas transferred to filters and hybridized with the JS21 LTR-specific probe (left) or with JS5 polytropic *env* probe (right). Note that the JS21 probe detects only one provirus (yielding two bands, each containing one LTR) in uninfected DNA, but also a large number of newly inserted proviruses, all of which correspond to new bands detectable with the polytropic *env* probe. Thus, all new proviruses seem to contain both alterations.

by a reduplication of the approximately 70-base region identified by functional, structural, and binding studies as containing the enhancer region with binding sites for at least six transcription factors (26). Thus reduplication apparently increases the transcriptional activity of the provirus (17, 18) and can affect the pathogenicity of the virus (17). To test whether this reduplication was also present in the *Bxv-1* provirus, we hybridized blots containing normal and tumor DNA with JS21 after digestion with a pair of restriction enzymes that yield a relatively short fragment extending from the beginning of U3 to the middle of R. An LTR with a reduplicated enhancer region is revealed by an increase in size of the fragment (Fig. 9). The endogenous *Bxv-1* provirus (which has not yet been cloned) contains a relatively short LTR, consistent with a single enhancer element, while all tumors contain a majority

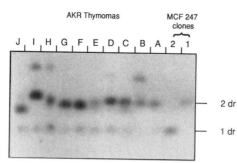

AKR Thymomas   MCF 247
                                                clones

J   I   H   G   F   E   D   C   B   A   2   1

— 2 dr

— 1 dr

Figure 9. Duplicated enhancer sequences in AKR tumors. The digestion was done with a pair of enzymes (*Pst*I and *Kpn*I) which yield a small fragment largely from the U3 region. Probe was with JS21. The two left lanes contained cloned DNA containing one or two copies of the enhancer region as markers. The remaining lanes contained the tumor DNAs shown in Fig. 8. Note that all have larger size LTRs and most have more than one size. dr, direct repeat(s).

of proviruses with reduplications (of various sizes) in this region. The interpretation that this difference in LTR size is due to reduplication of the enhancer region has been confirmed by cloning and sequencing several proviruses from the same tumors (data not shown). This analysis also revealed that all proviruses in a given tumor had identical *env* and LTR recombination points (which differed from tumor to tumor), but that the precise LTR duplication varied among proviruses in a single tumor, indicating that the duplication was a later event than the recombination.

The inferred series of events that gives rise to these viruses is shown in Fig. 10. Using our set of probes, we have performed a study of the time course of appearance of the virus recombinants in the thymus of AKR mice (not shown) which shows that through 10 weeks of age, only undetectable levels of new proviruses are present, but that new proviruses, all of which are MCF-like, increase rapidly between 10 and 16 weeks of age to a level greater than one provirus per cell in the entire thymus. Thus we conclude that the recombination and reduplication events are common in a mixed population of virus, which rapidly replicates until a large number of thymocytes are infected and one or more eventually become transformed.

CONCLUSIONS

The experiments we have presented here demonstrate the diversity of phenomena associated with the endogenous proviruses of mice. These proviruses are clearly relatively recent invaders of the mouse germ line, and most or all of them seem to be individually benign, although the insertion can (and has) caused a number of detectable and interesting mutations as well as handles for accessing the genes involved. The polymorphic nature of endogenous provirus insertion sites provides a very convenient tool for viewing the mouse genome in genetic experiments since a large number of genomic

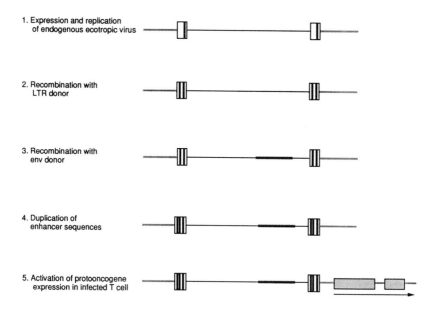

Figure 10. Probable events in AKR lymphoma. The progress of proviruses in the evolution of MCF virus is depicted schematically.

regions can be screened simultaneously. We are pursuing the applications of this technology along a number of lines.

The endogenous MLVs first came to light as mediators of "spontaneous" lymphoma. Clearly the process involves a well-defined series of recombination events between progeny of the various potential proviruses. It is highly unlikely that these events are in any way "directed"; rather, they are almost certainly strongly selected by their ability to replicate in the target organ. The replication advantage conferred by the LTR changes seems clear: it provides more suitable enhancer elements for T-cells. The role of the *env* substitution is less clear. It is possible that it has some chance stimulatory effect on T-cell growth, a factor that might promote both the ability of polytropic viruses to replicate in these cells as well as their oncogenicity. More direct experimentation on this point is clearly needed. Finally, the orchestrated series of recombination events provides an interesting model for in vivo change and evolution of retroviruses during a prolonged pathogenic process and may provide a useful model for thinking about selected genetic variation in other systems.

Acknowledgments. This project was supported by grants CA24530 and CA44385 from the National Cancer Institute. We thank S. Fenner for expert technical assistance and M. Bostic-Fitzgerald for manuscript preparation. W.N.F. is a Fellow of the Leukemia Society of America.

REFERENCES

1. **Chattopadhyay, S. K., M. W. Cloyd, D. L. Linemeyer, M. R. Lander, E. Rands, and D. R. Lowy.** 1982. Cellular origin and role of mink cell focus-forming viruses in murine thymic lymphomas. *Nature* (London) **295**:25–31.
2. **Coffin, J. M.** 1982. Endogenous retroviruses, p. 1109–1203. *In* R. Weiss, N. Teich, H. E. Varmus, and J. Coffin (ed.), *RNA Tumor Viruses*, vol. 1. Cold Spring Harbor Laboratory, Cold Spring Harbor, N.Y.
3. **Copeland, N. G., K. W. Hutchinson, and N. A. Jenkins.** 1983. Excision of the DBA ecotropic provirus in dilute coat-color revertants of mice occurs by homologous recombination involving the viral LTRs. *Cell* **33**:379–387.
4. **Crew, F. A. E., and L. Mirskaia.** 1931. The character "hairless" in the mouse. *J. Genet.* **25**:17–24.
5. **Evans, L. H., and M. W. Cloyd.** 1984. Generation of mink cell focus-forming viruses by Friend murine leukemia virus: recombination with specific endogenous proviral sequences. *J. Virol.* **49**:772–781.
6. **Frankel, W. N., J. M. Coffin, M. Cote, T. N. Seyfried, M. Rise, T. V. Rajan, F. K. Nelson, E. Selsing, and R. Gerstein.** 1991. Backcross data for endogenous *Mpmv*, *Pmv*, and *Xmv* proviruses. *Mouse Genome* **89**, in press.
7. **Frankel, W. N., J. P. Stoye, B. A. Taylor, and J. M. Coffin.** 1989. Genetic analysis of endogenous xenotropic murine leukemia viruses: association with two common mouse mutations and the viral restriction locus *Fv-1*. *J. Virol.* **63**:1763–1774.
8. **Frankel, W. N., J. P. Stoye, B. A. Taylor, and J. M. Coffin.** 1989. Genetic identification of endogenous polytropic proviruses by using recombinant inbred mice. *J. Virol.* **63**:3810–3821.
9. **Frankel, W. N., J. P. Stoye, B. A. Taylor, and J. M. Coffin.** 1990. A linkage map of endogenous murine leukemia proviruses. *Genetics* **124**:221–236.
10. **Furth, J., H. R. Siebold, and R. R. Rathbone.** 1933. Experimental studies on lymphomatosis of mice. *Am. J. Cancer* **19**:621–624.
11. **Gerstein, R. M., W. N. Frankel, C. Hsieh, J. H. Durdik, S. Rath, J. M. Coffin, A. Nisonoff, and E. Selsing.** 1990. Isotype switching of an immunoglobulin heavy chain transgene occurs by DNA recombination between different chromosomes. *Cell* **63**:537–548.
12. **Green, N., H. Hiai, J. H. Elder, R. A. Schwartz, R. H. Khiroya, C. Y. Thomas, P. N. Tsichlis, and J. M. Coffin.** 1980. Expression of leukemogenic recombinant viruses associated with a recessive gene in HRS/J mice. *J. Exp. Med.* **152**:249–264.
13. **Gross, L.** 1951. "Spontaneous" leukemia developing in C3H mice following inoculation, in infancy, with AK-leukemic-cell extracts, or AK-embryos. *Proc. Soc. Exp. Biol. Med.* **76**:27–32.
14. **Hartley, J. W., N. K. Wolford, L. J. Old, and W. P. Rowe.** 1977. A new class of murine leukemia virus associated with the development of spontaneous lymphomas. *Proc. Natl. Acad. Sci. USA* **74**:789–792.
15. **Hartung, S., R. Jaenisch, and M. Breindl.** 1986. Retrovirus insertion inactivates mouse $\alpha 1(I)$ collagen gene by blocking initiation of transcription. *Nature* (London) **320**:365–367.
16. **Hoggan, M. D., R. R. O'Neill, and C. A. Kozak.** 1986. Nonecotropic murine leukemia viruses in BALB/c and NFS/N mice: characterization of the BALB/c *Bxv-1* provirus and the single NFS endogenous xenotrope. *J. Virol.* **60**:980–986.
17. **Holland, C. A., P. Anklesaria, M. A. Sakakeeny, and J. S. Greenberger.** 1987. Enhancer sequences of a retroviral vector determine expression of a gene in multipotent hematopoietic progenitors and committed erythroid cells. *Proc. Natl. Acad. Sci. USA* **84**:8662–8666.

18. **Hollon, T., and F. K. Yoshimura.** 1989. Mapping of functional regions of murine retrovirus long terminal repeat enhancers: enhancer domains interact and are not independent in their contributions to enhancer activity. *J. Virol.* **63:**3353–3361.

19. **Jolicoeur, P., and E. Rassart.** 1980. Effect of *Fv-1* gene product on synthesis of linear and supercoiled viral DNA in cells infected with murine leukemia virus. *J. Virol.* **33:**183–195.

20. **Koch, W., W. Zimmerman, A. Oliff, and R. Friedrich.** 1984. Molecular analysis of the envelope and long terminal repeat of Friend mink cell focus-forming virus: implications for the functions of these sequences. *J. Virol.* **49:**828–840.

21. **Lung, M. L., J. W. Hartley, W. P. Rowe, and N. Hopkins.** 1983. Large RNase T1-resistant oligonucleotides encoding p15E and the U3 region of the long terminal repeat distinguish two biological classes of mink cell focus-forming type-C viruses of inbred mice. *J. Virol.* **45:**275–290.

22. **Morse, H.** 1981. The laboratory mouse: a historical perspective, p. 1–16. *In* H. L. Foster, J. D. Small, and J. G. Fox (ed.), *The Mouse in Biomedical Research*, vol. 1, *History, Genetics and Wild Mice*. Academic Press, Inc., New York.

23. **Nusse, R., and A. Berns.** 1988. Cellular oncogene activation by insertion of retroviral DNA: genes identified by provirus tagging, p. 95–119. *In* G. Klein (ed.), *Cellular Oncogene Activation*. Marcel Dekker, New York.

24. **O'Neill, R. R., A. S. Khan, M. D. Hoggan, J. W. Hartley, M. A. Martin, and R. Repaske.** 1986. Specific hybridization probes demonstrate fewer xenotropic than mink cell focus-forming murine leukemia virus *env*-related sequences in DNAs from inbred laboratory mice. *J. Virol.* **58:**359–366.

25. **Seperack, P. K., M. C. Strobel, D. J. Corrow, N. A. Jenkins, and N. G. Copeland.** 1988. Somatic and germ-line reverse mutation rates of the retrovirus-induced dilute coat-color mutation of DBA mice. *Proc. Natl. Acad. Sci. USA* **85:**189–192.

26. **Speck, N. A., and D. Baltimore.** 1987. Six distinct nuclear factors interact with the 75-base-pair region of the Moloney murine leukemia virus enhancer. *Mol. Cell. Biol.* **7:**1101–1110.

27. **Stoye, J. P., and J. M. Coffin.** 1985. Endogenous retroviruses, p. 357–404. *In* R. Weiss, N. Teich, H. E. Varmus, and J. M. Coffin (ed.), *RNA Tumor Viruses*, vol. 2. Cold Spring Harbor Laboratory, Cold Spring Harbor, N.Y.

28. **Stoye, J. P., and J. M. Coffin.** 1987. The four classes of endogenous murine leukemia virus: structural relationships and potential for recombination. *J. Virol.* **61:**2659–2669.

29. **Stoye, J. P., S. Fenner, G. E. Greenoak, C. Moran, and J. M. Coffin.** 1988. Role of endogenous retroviruses as mutagens: the hairless mutation of mice. *Cell* **54:**383–391.

30. **Teich, N., J. Wyke, and P. Kaplan.** 1985. Pathogenesis of retrovirus-induced disease, p. 187–248. *In* R. Weiss, N. Teich, H. E. Varmus, and J. M. Coffin (ed.), *RNA Tumor Viruses*, vol. 2. Cold Spring Harbor Laboratory, Cold Spring Harbor, N.Y.

31. **Teich, N., J. Wyke, T. Mak, A. Bernstein, and W. Hardy.** 1982. Pathogenesis of retrovirus-induced disease, p. 785–998. *In* R. Weiss, N. Teich, H. E. Varmus, and J. M. Coffin (ed.), *RNA Tumor Viruses*, vol. 1. Cold Spring Harbor Laboratory, Cold Spring Harbor, N.Y.

32. **Yang, W. K., J. O. Kiggins, D. M. Yang, C. Y. Ou, R. W. Tennant, A. Brown, and R. H. Bassin.** 1980. Synthesis and circularization of N- and B-tropic retroviral DNA in *Fv-1* permissive and restrictive mouse cells. *Proc. Natl. Acad. Sci. USA* **77:**2994–2998.

Viruses That Affect the Immune System
Edited by Hung Y. Fan et al.
© 1991 American Society for Microbiology, Washington, DC 20005

Chapter 13

Molecular Analysis of the Human T-Cell Leukemia Virus Type I *rex* Gene

Warner C. Greene, Sarah M. Hanly, Laurence T. Rimsky, and Yasmath F. Ahmed

Human T-cell leukemia virus type I (HTLV-I), a type C human retrovirus, has been etiologically linked with the adult T-cell leukemia (13, 26, 27, 41) and more recently with a progressive demyelinating syndrome termed HTLV-I-associated myelopathy or tropical spastic paraparesis (11, 25). Adult T-cell leukemia corresponds to an often aggressive and fatal tumor of CD4+ T-lymphocytes that occurs in regions of the world where HTLV-I infection is endemic. The general prognosis for patients with the acute form of this disease is poor, with few survivals observed beyond 6 months. A closely related human retrovirus, HTLV-II, has also been associated with certain T-cell tumors, specifically T-cell variants of hairy cell leukemia (7, 19). More recently, HTLV-II has been found to have spread widely within various intravenous drug abuse populations (20) but as yet has not been associated with a specific disease in this risk group.

Insights into the pathophysiological changes induced by these retroviruses have emerged with the molecular analysis of the HTLV-I provirus. Like other replication-competent retroviruses, the HTLVs encode functional *gag, pol,* and *env* gene products and contain long terminal repeats (LTRs) (34). However, unlike many acutely transforming animal retroviruses, HTLV-I and -II lack a classical oncogene. As well, these viruses do not consistently integrate adjacent to a cellular proto-oncogene (33). Rather, it seems likely

Warner C. Greene, Sarah M. Hanly, Laurence T. Rimsky, and Yasmath F. Ahmed • Howard Hughes Medical Institute, Department of Medicine, and Department of Microbiology and Immunology, Duke University Medical Center, Durham, North Carolina 27710.

that these viruses transform T cells via a novel mechanism importantly involving one or both of their *trans*-regulatory proteins Rex (p27[x-III]) and Tax (p40[x]). These proteins are encoded by a doubly spliced, polycistronic mRNA derived in part from the 3' pX region of the viral genome (31, 42). Rex and Tax are translated from different reading frames, and both appear obligately required for viral replication. The 40-kDa Tax protein serves as a transcriptional activator that augments activity of the HTLV-I LTR (18, 37) but also induces select cellular genes involved in T-cell growth. These genes include those encoding interleukin-2 and the alpha subunit of the interleukin-2 receptor (17, 40). In contrast, the 27-kDa Rex protein acts posttranscriptionally by inducing expression of the *gag, pol,* and *env* gene products necessary for the assembly of infectious virions (16, 18).

A considerable evolutionary gulf separates the HTLVs and human immunodeficiency viruses (HIVs). In terms of their disease manifestations, HIV type 1 (HIV-1) produces cytopathic rather than immortalizing effects within the human CD4+ subpopulation of T cells (4, 10, 32, 35). Despite these differences, HIV-1 similarly encodes nuclear *trans*-activating proteins termed Tat and Rev whose functions are broadly analogous to those of the Tax and Rex proteins of the HTLV-I and -II viruses (for review, see reference 8). In this regard, we have previously shown that the HTLV-I Rex protein can replace the activity of the HIV-1 Rev protein, as evidenced by Rex rescue of the replication of HIV-1 proviruses mutated in the *rev* gene (28). Thus, while the Rex and Rev proteins lack primary nucleotide or amino acid homology, the capacity of Rex to substitute for Rev highlights the apparent evolutionary convergence of HTLV-I and HIV-1.

HTLV-I ENVELOPE PROTEIN EXPRESSION

To dissect the effects of Rex on HTLV-I structural gene expression, we prepared several different HTLV-I reporter vectors. The HTLV-I genomic inserts contained within each of these constructs were promoted by the immediate early region of the cytomegalovirus. The pgTAX plasmid contains both of the two *tax* coding exons and the intervening intron that corresponds to the overlapping *env* gene. Production of the Rex protein, which is translated from the same transcript as Tax, was eliminated in this vector by the introduction of a mutation within the initiator methionine for Rex. This mutation thus allows for the cotransfection and testing in *trans* of wild-type or mutant forms of Rex cDNA. The pgTAX-LTR vector contains identical coding information but additionally contains the entire HTLV-I 3' LTR, which spans the putative Rex response element (RexRE). Fully spliced mRNAs produced by these vectors encode the 40-kDa Tax protein, while unspliced transcripts are translated to yield the 62-kDa HTLV-I envelope precursor protein.

Immunoprecipitation analyses of metabolically labeled, transiently transfected COS cell extracts revealed that the pgTAX expression vector lacking the RexRE expressed only fully spliced mRNAs encoding the Tax protein, both in the presence and absence of the HTLV-I Rex protein (14, 29). In contrast, the pgTAX-LTR vector, containing the RexRE, encoded envelope protein when Rex was provided in *trans*, but not in its absence. Of note, the HIV-1 Rev protein failed to activate *env* gene expression by this vector, suggesting that Rev cannot replace Rex in the HTLV-I system. The pgTAX-LTR vector system thus provides a sensitive and reproducible assay to measure Rex biological activity.

HTLV-II Rex AND HIV-1 Rev PROTEIN FUNCTIONS

Immunoprecipitation analysis of cells cotransfected with pgTAX-LTR and either an HTLV-II Rex expression plasmid or an HIV-1 Rev expression plasmid revealed that Rex-II but not Rev resulted in Env protein production. To verify that the 3' LTR of HTLV-I is sufficient alone to confer Rex-I and Rex-II responsiveness, we examined whether this *cis* element allowed Rex responsiveness in a heterologous HIV-1 vector system. This HIV-1 vector corresponded to a genomic *tat* expression vector containing the two *tat* coding exons and an intervening intron encompassing a part of the coding region of the *env* gene. Within the *env* gene resides the *cis*-regulatory sequence referred to as the Rev response element (RevRE) which is required for HIV-1 Rev action (30). In the pgΔTAT expression vector, the RevRE was deleted, thus rendering this expression vector refractory to both HIV-1 Rev and HTLV-I and -II Rex action. In turn, sequences corresponding to the RexRE were inserted in place of the RevRE. In the absence of Rex-I, pgΔTAT/RexRE vector expressed only the full-length 86-amino-acid (aa) form of the Tat protein, corresponding to the expression of fully spliced *tat* mRNAs in the cytoplasm. However, cotransfection of this vector with either the Rex-I or Rex-II expression vector resulted in the production of a truncated 72-aa form of the HIV-1 Tat protein, indicating the cytoplasmic expression and translation of unspliced *tat* mRNAs (22). This truncated 72-aa form of Tat protein is generated by the termination of translation immediately downstream of the first *tat* coding exon due to the presence of highly conserved stop codon. This stop codon is only represented in the unspliced mRNA species derived from this vector. As observed in the HTLV-I pgTAX-LTR construct, HIV-1 Rev was unable to induce the synthesis of the truncated 72-aa Tat protein from the pgΔTAT/RexRE plasmid. Thus, the RexRE alone is sufficient to convert

Rex-I and Rex-II responsiveness in a heterologous background but fails to confer HIV-1 Rev responsiveness.

CHARACTERIZATION OF THE HTLV-I RexRE

Like the HIV-1 RevRE (23), the HTLV-I RexRE is remarkable for its predicted stem-loop structure composed of 255 ribonucleotides (14). Interestingly, this RexRE largely resides within the long R region which uniquely distinguishes the HTLV-I family of viruses. This HTLV-I R region lies between the AAUAAA polyadenylation signal present in the U3 segment and the GU boxes located in the U5 region. These two recognition elements must be spatially approximated to direct efficient 3′ cleavage and polyadenylation of primary viral transcripts (reviewed by Birnstiel et al. [5]). Thus, the in vivo existence of the RexRE as a complex secondary structure is strongly supported by the fact that looping out of the large R region must occur for accurate 3′ mRNA processing. In fact, we have recently demonstrated a Rex-independent role for the RexRE in polyadenylation, using site-directed mutation to disrupt sequence of the primary stem within the RexRE. This mutation markedly attenuated polyadenylation in vivo and disrupted the stable and cooperative binding of polyadenylation factor 2 and cleavage factor 1 in vitro. The cooperative nature of the interaction of these factors was restored by compensatory mutations that restored the secondary RNA structure of the RexRE (1). This critical role of the Rex response element in effective viral polyadenylation raised the possibility that the action of Rex might involve this posttranscriptional process. However, as noted above, transfer of the 255-nucleotide RexRE to an intronic position with concomitant mutation of the AAUAAA hexamer motif did not alter the capacity of this element to confer Rex responsiveness. Thus, Rex responsiveness mediated through the RexRE and the role of the RexRE in polyadenylation are distinct, separable phenomena.

Additional studies of the RexRE have revealed that this element only functions in the sense orientation, further suggesting that it acts at the level of RNA rather than DNA (14). Consistent with this notion, Rex responsiveness was lost when the RexRE was positioned downstream of the polyadenylation site and thus not represented in the mRNA. More recent deletional studies have emphasized the important nature of the secondary structure of the RexRE to Rex action as well as the requirement for a small subregion of this element that likely forms a binding site for Rex or a cellular RNA binding protein (2).

THE Rex PROTEIN ENHANCES CYTOPLASMIC EXPRESSION OF INCOMPLETELY SPLICED VIRAL mRNAs

The precise mechanism by which Rex induces the expression of the HTLV-I structural proteins remains unclear. However, analysis of viral RNA within the nucleus of infected or transfected cells has revealed both spliced and unspliced viral mRNA species in the presence or absence of the Rex protein (14). In contrast, cytoplasmic expression of the unspliced or singly spliced family of viral mRNAs is only detectable in the presence of Rex. These findings suggest that the Rex either promotes disassembly of spliceosomes, thus allowing RNA transport of these unspliced viral mRNAs by default (see reference 6 for analysis of HIV-1 Rev), or, alternatively, Rex specifically activates cytoplasmic transport of these "nuclear sequestered" viral mRNAs.

MUTATIONAL STUDIES OF THE HTLV-I Rex PROTEIN

To study the mechanism of Rex action and to identify important peptide domains within this viral transactivator, a series of point mutations were introduced throughout the linear sequence of Rex (29). When analyzed in the pgTAX-LTR assay, three distinct types of mutants were identified. One class of mutants, corresponding to substitutions at aa 59–60 (M6), 64–65 (M7), and 119–121 (M13), not only lacked Rex biological activity but also functioned as *trans*-dominant repressors of the wild-type Rex protein. A second class of mutants containing substitutions at aa 5–7 (M1) and 14–15 (M2) similarly lacked biological activity but functioned in a recessive negative manner as they failed to impair wild-type Rex action. Finally, a third Rex mutant, altered at residues 143–145 (M15), displayed a partial loss of biological activity and lacked dominant negative properties. Twelve other point mutations exhibited full Rex activity as measured in this assay system.

In situ immunofluorescence staining of COS cells transiently transfected with vectors encoding the wild-type Rex protein revealed that this protein is targeted to the nucleus and in particular to the nucleoli of expressing cells (29, 36). Like the wild-type Rex protein, the *trans*-dominant Rex mutants (M6, M7, and M13) were similarly expressed in the nucleolar and nuclear compartments. In sharp contrast, the recessive negative mutants, M1 and M2, were not detectable in the nucleus. Rather, M1 was exclusively found in the cytoplasm while M2 was expressed in a homologous manner throughout the cell. Interestingly, the partially active M15 mutant was present in the nucleus but appeared specifically excluded from the nucleolus.

The *trans*-dominant Rex mutants (M6, M7, M13) were appropriately expressed in the nuclear compartment and inhibited the action of the wild-type Rex protein in a dose-related manner. Complete inhibition was routinely obtained at a 10-fold excess of each mutant over wild-type Rex. Since the HTLV-I Rex protein was able to replace the function of the HIV-1 Rev protein, we used the pgTAT assay to examine the capacity of the *trans*-dominant Rex mutants to block the function of Rev in the HIV-1 system. As discussed above, this assay exploits the capacity of both Rev and Rex to induce the expression of a truncated (72-aa) rather than full-length (86-aa) form of the Tat protein via the RevRE. At a 10-fold excess, the M6, M7, and M13 mutants each completely inhibited the biological effects of Rev (29). Furthermore, these Rex *trans*-dominant mutants also blocked the replication of HIV-1 proviruses. Specifically, the M6, M7, and M13 Rex mutants produced dose-related inhibition of p24 *gag* production in COS cells cotransfected with wild-type *rex* and a *rev*-deficient HIV-1 provirus. In contrast, the recessive negative Rex mutant, M1, lacked such inhibitory activity.

These findings with the various Rex mutants suggest the presence of at least two functional peptide domains within the Rex protein. One domain, defined by the M1 and M2 recessive negative mutations, is located at the N terminus and is apparently involved in targeting of this viral protein to the nucleus and nucleolus. In this regard, the M1 and M2 mutations alter residues within a positively charged segment previously implicated in nucleolar localization (36). It is also possible that this positively charged domain may be involved in the direct or indirect interactions of Rex with the putative subregion binding site identified in the RexRE. At present, however, no studies have conclusively shown that Rex binds directly to RNA. However, this possibility is strengthened by recent reports that Rev binds to the RevRE in a sequence-specific manner (9, 43). A second functional domain encompasses those residues altered in the M6, M7, and M13 mutations. These Rex mutant proteins both lack biological activity and display *trans*-dominant repressor properties. Interestingly, 54 aa separates M7 and M13 in the linear sequence of Rex, suggesting that the interaction of these discrete regions may require appropriate protein folding. The domain defined by these mutations likely plays an important role in the effector function of the Rex protein, which serves to regulate HTLV-I structural gene expression. At present, the molecular basis for inhibition by these *trans*-dominant mutants remains unclear. A number of potential mechanisms however exist, two of which have biological precedent (15). One possibility, exemplified by the I^{-d} mutants of *lac* repressor protein, is that mutant Rex proteins form mixed dimers or multimers with the wild Rex protein subunits and thereby inhibit the function of the wild-type Rex protein (24). However, it remains unknown whether Rex functions as a monomer or a more highly ordered structure in vivo. An attractive alternative hypothesis is that these mutations alter an activation domain that is required for function of the protein after binding.

Notwithstanding these uncertainties, the *trans*-dominant repressors of the HTLV-I *rex* gene join a small but expanding group of dominant negative viral proteins that hold future promise as an exciting new class of antiviral reagents. For example, *trans*-dominant mutants of the VP-16 protein of the herpes simplex virus (38), the Tax protein of HTLV-II (39), and the Rev and the Tat protein of HIV-I (12, 21) have now been described. If an effective and safe delivery system can be developed, these mutant genes or their corresponding proteins may find application as "intracellular immunogens" (3) to block the replication of their respective virus by inhibiting the essential function of the wild-type counterpart protein. In the case of the HTLV-I Rex *trans*-dominant mutants, these proteins are uniquely active in both the HTLV-I and HIV-1 systems and thus theoretically could afford protection against both of these pathogens.

CONCLUSIONS

The HTLV-I Rex protein plays an essential role in the life cycle of HTLV-I by activating the cytoplasmic expression of the unspliced or singly spliced viral transcripts in the cytoplasm that encode the viral structural proteins. The subsequent expression of these structural and enzymatic proteins, including Gag, Env, and reverse transcriptase, forms a prerequisite for virion assembly, budding, and further dissemination of the infection to new target cells. These actions of Rex are sequence specific, requiring the presence of a Rex response element that corresponds to a 255-nucleotide RNA stem-loop structure located within the 3' LTR. Mutational studies of the RexRE have revealed that its intrinsic secondary structure, as well as the presence of a localized subregion which likely forms the binding site for Rex or cellular RNA binding protein, is needed for Rex action. Independent of Rex, the RexRE also critically mediates the spatial approximation of the regulatory sequences needed for effective 3' mRNA cleavage and polyadenylation. The Rex protein contains at least two discrete structural domains. One domain is positively charged and involved in nuclear/nucleolar targeting. As well, this domain may mediate RNA binding of Rex or its interaction with a cellular RNA binding protein. Mutations within this domain lead to recessive negative Rex proteins. A second domain, defined by three different point mutations, appears to identify an effector or activation domain. These mutants not only lack intrinsic biologic activity but also function as dominant repressors of the wild-type Rex protein and HIV-1 Rev protein. The future development of safe and effective gene transfer techniques may allow for the application of these dominant negative mutant viral genes as novel antiviral therapeutics.

REFERENCES

1. **Ahmed, Y. F., G. M. Gilmartin, S. M. Hanly, J. R. Nevins, and W. C. Greene.** 1991. The HTLV-I Rex response element mediates a novel form of mRNA polyadenylation. *Cell* **64**:727–737.
2. **Ahmed, Y. F., S. M. Hanly, M. H. Malim, B. R. Cullen, and W. C. Greene.** 1990. Structure-function analysis of the HTLV-I Rex and HIV-1 Rev RNA response elements: insights into the mechanism of Rex and Rev action. *Genes Dev.* **4**:1014–1022.
3. **Baltimore, D.** 1988. Gene therapy. Intracellular immunization. *Nature* (London) **335**:395–396.
4. **Barré-Sinoussi, F., J. C. Chermann, F. Rey, M. T. Nugeyre, S. Chamarel, T. Gruest, C. Dauguet, C. Axler-Blin, F. Vezin-Brun, C. Rouzioux, W. Rozenbaum, and L. Montagnier.** 1983. Isolation of a T-lymphocyte retrovirus from a patient at risk for acquired immune deficiency syndrome. *Science* **220**:868–870.
5. **Birnstiel, M. L., M. Busslinger, and K. Strub.** 1985. Transcription termination and 3' processing: the end is in site! *Cell* **41**:349–359.
6. **Chang, D., and P. A. Sharp.** 1989. Regulation of HIV Rev depends upon recognition of splice sites. *Cell* **59**:789–795.
7. **Chen, I. S. Y., J. McLaughlin, J. C. Gasson, S. C. Clarke, and D. W. Golde.** 1983. Molecular characterization of genome of a novel human T-cell leukemia virus. *Nature* (London) **305**:502–505.
8. **Cullen, B. R., and W. C. Greene.** 1989. Regulatory pathways governing HIV-1 replication. *Cell* **58**:423–426.
9. **Daly, T. J., K. S. Cook, G. S. Gray, T. E. Maione, and J. R. Rusche.** 1989. Specific binding of HIV-1 recombinant Rev protein to the Rev responsive element *in vitro. Nature* (London) **342**:816–819.
10. **Gallo, R. C., S. Z. Salahuddin, M. Popovic, G. M. Shearer, M. Kaplan, B. F. Haynes, T. J. Palker, R. Redfield, J. Oleske, B. Sasai, G. Shite, P. Foster, and P. D. Markham.** 1984. Frequent detection and isolation of cytopathic retroviruses (HTLV-III) from patients with AIDS and at risk for AIDS. *Science* **224**:500–503.
11. **Gessain, A., F. Barin, J. C. Vernant, O. Gout, L. Maurs, A. Calender, and G. de The.** 1985. Antibodies to human T-lymphotropic virus type-I in patients with tropical spastic paraparesis. *Lancet* **ii**:407–410.
12. **Green, M., M. Ishimo, and P M. Loewenstein.** 1989. Mutational analysis of HIV-1 tat minimal domain peptides: identification of *trans*-dominant mutants that suppress HIV-LTR-driven gene expression. *Cell* **58**:215–223.
13. **Hanaoka, M., K. Takatsuki, and M. Shimoyama.** 1982. Adult T-cell leukemia and related disease. *Gan Monogr. Cancer Res.* **28**:1–237.
14. **Hanly, S. M., L. T. Rimsky, M. H. Malim, J. Kim, J. Hauber, M. Duc Dodon, S.-Y. Le, J. V. Maizel, B. R. Cullen, and W. C. Greene.** 1989. Comparative analysis of the HTLV-I Rex and HIV-1 Rev trans-regulatory proteins and their RNA response elements. *Genes Dev.* **3**:1534–1544.
15. **Herskowitz, I.** 1987. Functional inactivation of genes by dominant negative mutations. *Nature* (London) **329**:219–222.
16. **Hidaka, M., J. Inoue, M. Yoshida, and M. Seiki.** 1988. Post-transcriptional regulator (*rex*) of HTLV-I initiates expression of viral structural proteins but suppresses expression of regulatory proteins. *EMBO J.* **7**:519–523.

17. **Inoue, J., M. Seiki, T. Taniguchi, S. Tsuru, and M. Yoshida.** 1986. Induction of interleukin-2 receptor gene expression by p40^x encoded by human T-cell leukemia virus type I. *EMBO J.* **5**:2883–2888.

18. **Inoue, J.-I., M. Yoshida, and M. Seiki.** 1987. Transcriptional (p40^x) and post-transcriptional (p27^{x-III}) regulators are required for the expression and replication of human T-cell leukemia virus type I genes. *Proc. Natl. Acad. Sci. USA* **84**:3653–3657.

19. **Kalyanaraman, V. S., M. G. Sarngadharan, M. Robert-Guroff, I. Miyoshi, D. Blayney, D. Golde, and R. C. Gallo.** 1982. A new subtype of human T-cell leukemia virus (HTLV-II) associated with a T-cell variant of hairy cell leukemia. *Science* **218**:571–573.

20. **Lee, H., P. Swanson, V. S. Shorty, J. A. Zack, J. D. Rosenblatt, and I. S. Y. Chen.** 1989. High rate of HTLV-II infection in seropositive IV drug abusers in New Orleans. *Science* **244**:471–475.

21. **Malim, M., S. Böhnlein, J. Hauber, and B. R. Cullen.** 1989. Functional dissection of the HIV-1 rev *trans*-activator—derivation of a *trans*-dominant repressor of Rev function. *Cell* **58**:205–214.

22. **Malim, M. H., J. Hauber, R. Fenrick, and B. R. Cullen.** 1988. Immunodeficiency virus rev *trans*-activator modulates the expression of the viral regulatory genes. *Nature* (London) **335**:181–183.

23. **Malim, M. H., J. Hauber, S.-Y. Le, J. V. Maizel, and B. R. Cullen.** 1989. The HIV-1 *rev trans*-activator acts through a structured target sequence to activate nuclear export of unspliced viral mRNA. *Nature* (London) **338**:254–257.

24. **Miller, J. H.** 1980. The lacI gene: its role in lac operon control and its use as a genetic system, p. 31–88. *In* J. H. Miller and W. S. Reznikoff (ed.), *The Operon*, 2nd ed. Cold Spring Harbor Laboratory, Cold Spring Harbor, N.Y.

25. **Osame, M., K. Usuku, N. Ijichi, H. Amitani, A. Igata, M. Matsumoto, and H. Tara.** 1986. HTLV-I associated myelopathy, a new clinical entity. *Lancet* **i**:1031–1032.

26. **Poiesz, B. J., F. W. Ruscetti, A. F. Gazdar, P. A. Bunn, J. D. Minna, and R. C. Gallo.** 1980. Detection and isolation of type C retrovirus particles from fresh and cultured lymphocytes of a patient with cutaneous T-cell lymphoma. *Proc. Natl. Acad. Sci. USA* **77**:7415–7419.

27. **Poiesz, B. J., F. W. Ruscetti, M. S. Reitz, V. S. Kalyanaraman, and R. C. Gallo.** 1981. Isolation of new type C retrovirus (HTLV) in primary uncultured cells of a patient with Sezary T-cell leukemia. *Nature* (London) **294**:268–271.

28. **Rimsky, L., J. Hauber, M. Dukovich, M. H. Malim, A. Langlois, B. R. Cullen, and W. C. Greene.** 1988. The *rex* protein of HTLV-I can functionally replace the *rev* protein of HIV-1. *Nature* (London) **335**:738–740.

29. **Rimsky, L. T., M. Duc Dodon, E. P. Dixon, and W. C. Greene.** 1989. Mutational analysis of the HTLV-I Rex transactivator: transdominant inactivation of HTLV-I and HIV-1 gene expression. *Nature* (London) **341**:453–456.

30. **Rosen, C. A., E. Terwilliger, A. Dayton, J. G. Sodroski, and W. A. Haseltine.** 1988. Intragenic *cis*-acting *art* gene-responsive sequences of the human immunodeficiency virus. *Proc. Natl. Acad. Sci. USA* **85**:2071–2075.

31. **Rosenblatt, J. D., A. J. Cann, D. W. Golde, and I. S. Y. Chen.** 1986. Structure and function of the human T-cell leukemia virus II genome. *Cancer Rev.* **1**:115.

32. **Sarngadharan, M. G., M. Popovic, L. Bruch, J. Schupbach, and R. C. Gallo.** 1984. Antibodies reactive with human T-lymphotropic retroviruses (HTLV-III) in the serum of patients with AIDS. *Science* **224**:506–508.

33. **Seiki, M., R. Eddy, T. Shows, and M. Yoshida.** 1984. Non-specific integration of HTLV provirus genome into adult T-cell leukemia cells. *Nature* (London) **309**:640–642.

34. **Seiki, M., S. Hattori, Y. Hirayama, and M. Yoshida.** 1983. Human adult T-cell leukemia virus: complete nucleotide sequence of the provirus genome integrated in leukemia cell DNA. *Proc. Natl. Acad. Sci. USA* **80**:3618–3622.

35. **Siegal, F. P., C. Lopez, G. S. Hammer, A. E. Brown, S. J. Komfeld, J. Gold, J. Hassett, S. Z. Hirschman, S. Cunningham-Rundles, and D. Armstrong.** 1981. Severe acquired immunodeficiency in male homosexuals, manifested by chronic perianal ulcerative herpes simplex lesions. *N. Engl. J. Med.* **305:**1439–1444.

36. **Siomi, H., H. Shida, S. Hyun Nam, T. Nosaka, M. Maki, and M. Hatanaka.** 1988. Sequence requirements for nucleolar localization of human T-cell leukemia virus type I pX protein which regulates viral RNA processing. *Cell* **55:**197–209.

37. **Sodroski, J. G., C. A. Rosen, and W. A. Haseltine.** 1984. Transacting transcriptional activation of the long terminal repeat of human T lymphotropic viruses in infected cells. *Science* **225:**381–385.

38. **Triezenberg, S. J., R. C. Kinsbury, and S. L. McKnight.** 1988. Functional dissection of VP16, the *trans*-activator of herpes simplex virus immediate early gene expression. *Genes Dev.* **2:**718–729.

39. **Wachsman, W., A. J. Cann, J. L. Williams, D. J. Slamon, L. Souze, N. P. Shah, and I. S. Chen.** 1987. HTLV-*x* gene mutants exhibit novel transcriptional regulatory phenotypes. *Science* **235:**674–677.

40. **Wano, Y., M. Feinberg, J. B. Hosking, H. Bogerd, and W. C. Greene.** 1988. Stable expression of the HTLV-I Tax protein specifically activates cellular genes involved in human T cell growth. *Proc. Natl. Acad. Sci. USA* **85:**9733–9737.

41. **Yoshida, M., I. Miyoshi, and Y. Hinuma.** 1982. Isolation and characterization of retrovirus from cell lines of a human adult T-cell leukemia and its implication of the disease. *Proc. Natl. Acad. Sci. USA* **79:**2031–2035.

42. **Yoshida, M., and M. Seiki.** 1987. Recent advances in the molecular biology of HTLV-I: *trans*-activation of viral and cellular genes. *Annu. Rev. Immunol.* **5:**541–559.

43. **Zapp, M. L., and M. R. Green.** 1989. Sequence-specific RNA binding by the HIV-1 Rev protein. *Nature* (London) **342:**714–716.

Part IV

ONCOGENESIS BY HERPESVIRUSES

Viruses That Affect the Immune System
Edited by Hung Y. Fan et al.
© 1991 American Society for Microbiology, Washington, DC 20005

Chapter 14

Immortalization of Human B-Lymphocytes by Epstein-Barr Virus

Bill Sugden

Epstein-Barr virus (EBV) is a human herpesvirus that causes infectious mononucleosis and is causally associated with several lymphoid tumors and one epithelial tumor in people. Both in infectious mononucleosis and in the lymphoid tumors with which EBV is associated, the virus has been found to have infected B-lymphocytes and to maintain its DNA in those cells as a plasmid. These infected B cells, upon explanting in cell culture, proliferate ad infinitum and are said to be immortalized. In vitro EBV infects primary human B-lymphocytes and also immortalizes them. It seems likely that the contribution of EBV to the B-lymphoid diseases with which it is associated is the immortalization of the infected cells. This review has three objectives: first, to recount the evidence that immortalization of B-lymphocytes by EBV in vitro is efficient; second, to relate the circumstantial evidence that indicates that several genes of EBV are likely to be required for this immortalization event; and third, to describe our current understanding of three genes of EBV that are likely, or known, to be required for immortalization.

IMMORTALIZATION OF HUMAN B-LYMPHOCYTES BY EBV IS EFFICIENT

A variety of experiments in the 1970s established that EBV infects human B-lymphocytes efficiently (19, 22, 36, 51). Human peripheral mononuclear

Bill Sugden • McArdle Laboratory for Cancer Research, University of Wisconsin, Madison, Wisconsin 53706.

cells were separated by various means, and the B-lymphocyte fraction was shown to be susceptible to infection by EBV (22, 36). The number of B-lymphocytes infected could be monitored by scoring for nuclear antigens encoded by EBV in these cells, and it was found that 50% or more of peripheral B-lymphocytes from adult donors could be infected by EBV (43). Clonal immortalization assays which used either limiting dilution or plating in semisolid media were employed then to enumerate the number of infected cells that were immortalized. Initially this approach was used with total peripheral lymphocytes, and more recently it has been extended to purified B-cell fractions. In these initial experiments, approximately 1 cell per 1,000 lymphocytes grew out immortalized (19, 51). The bulk lymphocytes used for the initial transformation assays contained between 10 and 20% B-lymphocytes. Thus, approximately 1 cell per 100 B-lymphocytes is immortalized by EBV. This number, however, was not corrected for the cloning efficiencies of those cells upon infection. The immortalized cells were therefore tested for their cloning efficiency by the same kind of clonal assay. In so doing, it was found that these cells, upon retesting, had cloning efficiencies on the order of 1 to 10% (51, 52). If we assume that the cloning efficiency of these cells upon infection is equal to what is observed many weeks after the cells have been immortalized, that is, on the order of 1 to 10%, then the 1 cell per 100 that grew out to be immortalized actually represents between 10 and 100% of the infected B-lymphocytes being immortalized. These calculations indicate that EBV efficiently immortalizes the human peripheral B cells that it infects.

There is one important consequence of this efficient immortalization of human B-lymphocytes by EBV. Because a large fraction of infected cells are capable of being immortalized, although their cloning efficiency remains only 1 to 10%, it is clear that a normal B cell in the peripheral blood can be immortalized. In other words, we need not look for variants among the primary B-lymphocytes in our peripheral blood that are competent to be immortalized after infection by EBV. Rather, mere infection by this virus yields descendants that are capable of indefinite growth in vitro.

Another kind of experiment also performed in the late 1970s indicated that the specific immortalizing capacity of Epstein-Barr virions is high. The total number of virions in a stock of EBV was first assayed by measuring the amount of EBV DNA present, and then the stock of virus was used to immortalize peripheral B-lymphocytes. It was found that approximately 1 particle in 30 of the DNA-containing particles was capable of immortalizing human B-lymphocytes (51). This high specific immortalizing capacity of EBV virions is striking when contrasted with the inefficient ability of simian virus 40 (SV40) to transform mouse cells in culture. It has been found that optimal transformation by SV40 requires on the order of 100,000 virus particles per cell (42).

That approximately 1 in 30 DNA-containing particles of EBV is competent to immortalize B-lymphocytes has an important consequence. It indicates that it is a wild-type Epstein-Barr virion that is capable of immortalizing what we now know to be a normal human B-lymphocyte. We need not look for variants or mutants of EBV being required for the immortalization event. This interpretation is in contrast to our understanding of transformation of established rodent cells by SV40, for example. Often mutants of SV40 are maintained in the transformed cell and presumably contribute to the transformation event (32, 49).

MULTIPLE EBV GENES ARE LIKELY TO BE REQUIRED FOR EFFICIENT IMMORTALIZATION

Although EBV immortalizes human B-lymphocytes efficiently, it appears that EBV is likely to require several viral genes in order to immortalize cells, unlike SV40, which requires only a single gene. There are two kinds of circumstantial evidence that support this likelihood. The first is derived from a vast amount of work done by several groups which have had as their goal the identification of viral genes that are expressed in immortalized B-lymphoblasts (reviewed in references 13, 24, 29, 34, and 48). Historically, the protein products of these viral genes were first detected with human antisera from patients with Burkitt's lymphoma, one of the lymphoid tumors associated with infection by EBV. The protein products expressed from the viral RNAs now are often detected with antibodies generated in rabbits against fusion proteins produced in *Escherichia coli*. Viral genes expressed in immortalized cells have also been identified by detecting their RNAs and analyzing cDNA copies of these RNAs in detail. The sum of all of this research indicates that on the order of 10 viral genes are found usually to be expressed in a variety of cells immortalized by EBV in vitro (reviewed in references 13, 24, 29, 34, and 48). Three of these viral genes encode proteins which are in the plasma membrane. Seven of them encode proteins which are found in the nucleus and are referred to as EBV nuclear antigens (EBNAs). In addition to these 10 viral genes that express protein products, two small viral RNAs, referred to as the EBERs, are expressed in immortalized cells (2, 21, 30, 44). The EBERs are presumably transcribed by RNA polymerase III and may be important for the immortalization process (2, 21).

It seems likely that the function of many of the viral genes that are expressed in the immortalized cell is to maintain the immortalized state. There are, however, two caveats to keep in mind in considering this likelihood. First, it is possible that some of the genes that EBV expresses in the immortalized B-lymphoblast play little or no role in immortalization in B-lymphocytes, but rather are important when EBV infects and is maintained in epithelial cells, the other known host cell for infection by EBV in human beings (12, 27). The second caveat to recall is that not all of the genes of the EBV that have been found to be expressed in some immortalized B-lymphoblasts have been examined for their expression in cells recently immortalized in vitro. The phenotypes of cells immortalized by EBV do change upon propagation in cell culture. It is reasonable, therefore, to consider only those genes of EBV that are consistently expressed in recently immortalized cells as being candidates for contributing to the immortalized state.

There is a second kind of evidence that indicates that multiple genes of EBV are likely to be required for immortalization of infected B-lymphocytes. In one study a frameshift mutagen, N-acetoxyacetylaminofluorene (N-acetoxy AAF), was used to determine the target size for the immortalizing functions of EBV. This target size was derived by inactivating EBV and herpes simplex virus type 1 (HSV-1) each with several concentrations of N-acetoxy AAF, comparing the dose response of the capacity of EBV to immortalize cells with that of HSV-1 to form a plaque, and normalizing EBV's target size to the size of HSV-1 DNA required to form a plaque. This study found that between 20 and 25% of the DNA of EBV is required to immortalize human B-lymphocytes (33). This finding supports the interpretation that several genes of EBV are likely to be required for it to immortalize B cells.

N-Acetoxy AAF not only is an alkylating agent, but also inhibits transcription (47). It is possible that the very long transcripts, which are now known to be the parents from which RNAs are spliced that encode EBV gene products (reviewed in reference 48), could themselves be inhibited by N-acetoxy AAF binding to EBV DNA. In other words, it is possible that this inactivating agent inhibits the wild-type expression of EBV genes not only because it eventually renders them mutant, but also because it reduces their transcription from DNA templates prior to the first round of replication of the templates. It is known, for example, that several genes of EBV are expressed in the immortalized cell from RNAs whose primary transcripts are on the order of 80 kb in length (6, 7). These considerations indicate that the large target size found for the immortalizing functions of EBV when N-acetoxy AAF was used as an inactivating agent may be overestimates of the coding capacity of the viral DNA needed for immortalization. Nevertheless, these experiments are consistent with the notion that several genes of EBV are required for immortalization.

If several genes of EBV are required to immortalize human B-lymphocytes, which ones are they? This question cannot now be answered. However, we know enough about at least three genes of EBV that are expressed in immortalized B-lymphoblasts to guess that each of them is likely to be required. Of the genes of EBV that are expressed in immortalized cells we know most about these three; it therefore remains possible that others of the expressed genes are likely to be required for immortalization too. The three genes that appear today to be likely candidates for being required for immortalization are EBNA-1, EBNA-2, and LMP (latent membrane protein). Collectively, these EBV genes are known to be required to mediate viral plasmid DNA replication, to affect the expression of viral and cellular genes, and to affect the growth characteristics of cells that express them. I shall review the current status of our understanding of each of these three viral genes to provide evidence that they are either now known to be or likely to be required for immortalization.

EBNA-1

EBNA-1 was the first gene of EBV to be found to be consistently expressed in EBV-immortalized cells (39) and to be mapped to the viral genome (54). It was also the first gene of EBV for which a function was identified (60). EBNA-1 was shown to be expressed in EBV-immortalized cells by using anticomplement immunofluorescence and human antisera (39). In fact, this assay for this antigen is often considered to be diagnostic for the presence of EBV in immortalized B-lymphoblasts. Clearly, its accuracy is a function of the specificity of the serum used. EBNA-1 was mapped to a particular open reading frame in the viral genome by transfecting fragments of EBV DNA into established rodent cell lines and showing that one specific open reading frame encoded an antigen that scored positively in the anticomplement immunofluorescent assay (54). A function of EBNA-1 was identified when the *cis*-acting element *ori*P, which is required for plasmid replication of EBV plasmids, was identified (59). Once the *cis*-acting sequence was identified, it became possible to search the viral genome for those genes required to permit plasmid replication. Eleven overlapping fragments of EBV DNA, which together constitute the entire viral genome, were individually introduced into human cells. These clones of cells, each of which carried a particular segment of EBV DNA, were then used as recipients to determine whether they could support the replication of a plasmid carrying *ori*P. Only one region of EBV DNA was shown to encode this function (60). This region is the open reading frame for the EBNA-1 protein (31). Only EBNA-1, therefore, is required in *trans* for *ori*P-mediated plasmid DNA replication. It has been shown that the carboxy-terminal one-third of EBNA-1 is a site-specific DNA binding protein and that the whole protein has this same activity (23, 38). EBNA-1 binds to the specific sequences of *ori*P which have been shown also to be required in *cis* for plasmid DNA replication (41). Thus, one function of

EBNA-1 is to permit plasmid DNA replication of the EBV genomes present in immortalized cells.

EBNA-1 probably has a second function. One of the two components of *ori*P to which EBNA-1 binds is referred to as the family of repeats. The family of repeats scores as a transcriptional enhancer in the presence of EBNA-1 for a variety of heterologous promoters (40). It also enhances expression from at least one viral promoter, the promoter located within the *Bam*HI C fragment of EBV (53). This promoter has been shown to lead to the expression of a variety of viral RNAs in the immortalized cell (8) which encode at least three viral gene products (6, 7). The experiments that indicate that EBNA-1 can activate the enhancing element within *ori*P to affect the promoter within *Bam*HI C have all used small plasmids (53). No test of the effect of EBNA-1 on transcription for the entire viral genome has been made. Such experiments today would be very difficult to envision. With the caveat that we must argue for a transcriptional effect of EBNA-1 on the entire viral genome from its known effects on its target enhancer in smaller plasmids, it seems likely that EBNA-1 will activate the transcription of some viral genes resident in the viral genome in immortalized cells.

It is certainly possible that EBNA-1 performs other functions than those involved in plasmid DNA replication, and perhaps enhancement of viral transcription. Recent observations made with newly established Burkitt's lymphoma cell lines, or from Burkitt's lymphoma biopsies themselves, indicate that EBNA-1 may perform some function which directly affects cell proliferation. It now appears that the only viral gene product that can be detected in some Burkitt's lymphoma biopsies or some of the cell lines established from these biopsies is EBNA-1 (45). If the viral genome in these cells is contributing to the immortalized phenotype of the Burkitt's lymphoma cells, then that contribution must be mediated by EBNA-1. If the viral genome is not contributing to the immortalized phenotype of the cells, it is not at all clear why the genomes are not lost from those cells. There is precedent for the loss of some viral genomes of one genotype from a cell that contains genomes of two genotypes. In particular, the Jijoye Burkitt's lymphoma cell line gave rise to a deleted variant now referred to as P3HR1 (20, 37). The parental cell presumably at one time had both the wild-type and the deleted variant and over time the wild-type genome was lost from these cells. With this precedent in mind, it seems likely that the EBV genomes in Burkitt's lymphoma cell lines that are recently derived from biopsies will be providing those cells some selective growth advantage. If this conjecture is true, then the selective growth advantage presumably is mediated by EBNA-1. How EBNA-1 might provide this advantage remains a mystery.

Given that EBNA-1 is required to mediate plasmid DNA replication of EBV genomes in immortalized cells, it seems evident that this viral gene product will be required for the efficient immortalization of B-lymphoblasts by EBV. We have as yet no compelling test of this likelihood. Furthermore, it remains possible that EBNA-1 will be required to initiate plasmid replication in cells after primary infection, but may not be needed to maintain viral DNA replication. Were EBNA-1 to have such a limited time of action, it still is likely that plasmid replication, per se, is required for efficient immortalization of B cells by EBV in that the vast majority of cells that have been immortalized in vitro, when studied, have been found to maintain their viral DNA as plasmids.

EBNA-2

It has been possible to establish that EBNA-2, another nuclear antigen encoded by EBV, is required for the immortalization of infected B-lymphocytes. This test grew out of three observations. First, it has been known for some time that one strain of EBV that is deleted both for EBNA-2 and part of another nuclear antigen encoded by EBV, referred to either as EBNA-LP or EBNA-5, is incapable of immortalizing infected B-lymphocytes (35). Either one or both of these affected viral genes, or perhaps undetected lesions elsewhere in the P3HR1 genome, could theoretically render it incapable of immortalizing cells. The second set of observations that contributed to the test for the requirement of EBNA-2 in immortalization identified means of inducing the lytic phase of the life cycle of EBV in selected EBV-immortalized cell lines (10, 61). The third set of observations which contributed to the genetic test was derived from studies of HSV-1 and EBV. The efficient recombination associated with HSV-1 in infected cells was found to depend both upon its lytic mode of replication and on the viral genes required for this replication (58). The *cis*-acting sequences of EBV required for its lytic mode of DNA replication have been identified (16). Vectors that contain the *cis*-acting sequences of EBV required for the lytic mode of its replication, termed *oriLyt*, when introduced into cells that are supporting the lytic phase of replication of endogenous EBV, replicate and recombine with the endogenous EBV (17). This recombination appears to be mediated by homologous DNA sequences between the vector and the endogenous viral DNA and parallels many of the observations made for homologous recombination with HSV-1. The confluence of these three sets of observations has permitted the development of vectors that carry either wild-type EBNA-2 and wild-type EBNA-LP or mutations in either one of these genes, along with *oriLyt*. These vectors can be introduced into the P3HR1 cell line deleted for part of EBNA-LP and all of EBNA-2; that cell line can be induced to support the lytic phase of EBV's life cycle and subsequently yield the release of recombinant, infectious viruses. These viruses infect primary B-lymphocytes, and only those that have a wild-type EBNA-2 are competent to immortalize cells (17). These experiments indicate that EBNA-2 is required to immortalize primary B-lymphocytes, and that EBNA-LP, although it may affect the efficiency of immortalization, is not itself required (17).

Inasmuch as EBNA-2 is required for EBV to immortalize human B-lymphocytes, it has become particularly desirable to identify activities of this viral gene product which are required for immortalization. Several activities of, or phenotypes induced by, EBNA-2 have already been identified. EBNA-2 expressed from a retroviral vector has been introduced and expressed in Rat-1 cells and yields cells that are less dependent upon serum for their growth (11) than are their parents. EBNA-2 expressed from a retroviral vector when introduced into Loukes cells, an EBV-negative Burkitt's lymphoma cell line, increases the expression of CD23 RNA in these cells (57). CD23 is a cell surface antigen whose expression at the cell surface is dramatically increased in B cells by infection with EBV (25).

The expression of this same activation marker is also increased upon expression of EBNA-2 in another type of cell. A variety of EBV-negative Burkitt's lymphoma cell lines have been infected with the P3HR1 strain of EBV, and derivatives have been selected that maintain viral DNA (9). Recall that P3HR1 is a strain of EBV that carries a deletion of part of the EBNA-LP open reading frame and of all of the EBNA-2 open reading frame. When EBNA-2 expressed from an *ori*P vector is introduced into these derivatives of Burkitt's lymphoma cell lines, the expression of the RNA of two activation markers, CD21 and CD23, is increased (9). CD21 is also known as CR2 and is thought to be a receptor for infection by EBV (15).

EBNA-2 has also been found to affect the sensitivity of several B-cell lines to the antiproliferative effect of alpha interferon. Derivatives of B-cell lines shown to express EBNA-2 have been found to be 100-fold more resistant to these antiproliferative effects than are the EBNA-2-negative parents (1).

Finally, the expression of EBNA-2 from a variety of vectors in EBV-negative Burkitt's lymphoma cell lines leads to an increased expression of the RNA of the proto-oncogene c-*fgr* (28). The c-*fgr* proto-oncogene is a member of the *src* proto-oncogene family. Its increased expression resulting from the expression of EBNA-2 could certainly contribute to the altered growth control in cells induced by infection with EBV.

Several activities have been assigned to the EBNA-2 gene of EBV when it is expressed in established human B-lymphoblastoid cell lines in the absence or in the presence of other viral genes. Some of these activities affect the level of accumulation of specific cellular RNAs and could function at the level of transcription, RNA processing, or RNA stability. It is now important to identify one or more of these activities that correlate with the activity of EBNA-2 in the immortalization of infected human B-lymphocytes. When such an activity is identified, its biochemical dissection should lead to an understanding of how EBNA-2 contributes to the efficient immortalization of B cells by EBV.

One puzzling observation about EBNA-2 needs to be considered in any attempt to explain its contribution to immortalization of cells by EBV. It appears that EBNA-2 under certain circumstances is disadvantageous to immortalized cells either in cell culture or in the infected human host. At least one variant of EBV has arisen in cell culture, the P3HR1 strain of EBV, which is derived from its parent Jijoye and which fails to express EBNA-2 (37). The Jijoye cell line contains the Jijoye strain of EBV, which has the EBNA-2 gene and expresses it, while its progeny cell, P3HR1, contains the P3HR1 strain of EBV which has this gene deleted. The cell containing the P3HR1 variant DNA must at some time have overgrown the Jijoye parent; this inference indicates that the absence of EBNA-2 can provide a selective advantage in cell culture. Similarly, some Burkitt's lymphoma biopsies fail to express EBNA-2 (45). Presumably, the cell that gave rise to these Burkitt's lymphoma cells at one time expressed EBNA-2. It is possible to posit that the reason the Burkitt's lymphoma biopsy cells no longer express EBNA-2 is that there is an immunological selection against its expression (26). However, no such explanation can be offered to explain why the P3HR1 cell line arose in the background of Jijoye cells in cell culture. It therefore seems likely that under certain circumstances the expression of EBNA-2 is disadvantageous both in cell culture and in vivo in the human host. Even if EBNA-2, under certain circumstances, is disadvantageous for a cell it is clear that the genetic experiments described above demonstrate that EBNA-2 itself is required for immortalization of human B-lymphocytes by EBV.

LMP

A third EBV gene likely to be involved in immortalization is termed LMP (latent membrane protein). Unlike EBNA-1 and EBNA-2, this protein is found primarily at the cell surface. It is not known whether the LMP gene is required for immortalization of cells by EBV, but the activities associated with it make it likely. In particular, LMP has been shown to be an oncogene in established rodent cells (3, 55). These cells can be rendered anchorage independent by an appropriate level of expression of LMP, and the resulting anchorage-independent clones are tumorigenic when tested in nude mice (3, 55). That LMP scores as an oncogene in these assays makes it likely, by analogy to other studied oncogenes, that LMP will contribute to the immortalization process mediated by EBV.

Several activities have been assigned to LMP expressed in EBV-negative B-lymphoblasts. It has been found to affect the expression of CD23, of LFA-

1, of ICAM-1, and of vimentin (5, 56). The accumulation of the RNA for these cellular genes is increased in cells that express LMP. The mechanism by which this increase is accomplished is not known.

The proposed structure of LMP within the plasma membrane is striking and may provide insight as to how LMP affects cellular growth control and cellular gene expression. This protein is thought to span the plasma membrane six times, with approximately 25 amino acids at its amino terminus being cytoplasmic and almost 200 amino acids at its carboxy terminus also being cytoplasmic (14). Its structure resembles that of an ion channel or a receptor (50). There are precedents for ion channels or receptors carrying out at least some of the activities that have already been assigned to LMP.

A large number of deletions in the LMP gene have been made which have been used to express a number of derivatives of LMP (4, 18). These derivatives have permitted the identification of regions of the protein that are required for its known activities. In particular, all of the six membrane-spanning domains and most, if not all, of the amino terminus appear to be required for any activity. However, none of the carboxy terminus is required for any of the activities of LMP tested to date (4, 18). These mutational analyses underscore the significance of the six membrane-spanning domains for activities of LMP. They have not, unfortunately, distinguished between the possibilities of LMP being an ion channel, a receptor, or something as yet undreamt of (46).

The study of the deleted variants of LMP also provides an additional and striking piece of information about this gene product. In particular, it has been observed that when the variants of LMP that render BALB/3T3 cells anchorage independent for growth are expressed in any of eight cell lines at high levels, these variants kill the cells (18). A similar observation has also been made for wild-type LMP (18). When variants that fail to yield anchorage-independent growth in BALB-3T3 cells are expressed at high levels in any of those eight cell lines, they do not affect the survival of these cell lines (18). It seems almost certain, therefore, that the activities LMP encodes that are required for anchorage-independent growth of BALB/3T3 cells are also lethal when expressed at high levels. These observations indicate that it is likely that the level of expression of LMP in immortalized cells needs to be controlled rigorously. Considerations of EBNA-2 discussed above indicate that its level of expression may well need to be tightly controlled, too. Such control may also be required for other genes of EBV that play a role in immortalization. It seems reasonable to speculate, therefore, that several of the EBV genes that are expressed in immortalized cells play important roles in regulating the expression of certain viral genes in order that their level of expression is consistent with optimal growth of the cells.

EBV immortalizes human B-lymphocytes efficiently, and it appears that several genes of the virus are probably required for this efficient immortalization. This complexity of viral gene expression seems at odds with the limited expression of transforming or immortalizing genes in other transforming viruses. However, the immortalization of B-cells by EBV may well be an event that is intrinsic to the life cycle of EBV in its human host. If this notion were to be correct, then the virus could afford to devote several genes to insure that immortalization of its infected cell is efficient.

Acknowledgments. I thank my colleagues Toni Gahn, Shikha Laloraya, Jennifer Martin, Tim Middleton, and Donata Oertel for help in revising this manuscript. I was supported by Public Health Service grants CA-22443 and CA-07175.

REFERENCES

1. **Aman, P., and A. von Gabain.** 1990. An Epstein-Barr virus immortalization associated gene segment interferes specifically with the IFN-induced anti-proliferative response in human B-lymphoid cell lines. *EMBO J.* **9:**147–152.
2. **Arrand, J. R., and L. Rymo.** 1982. Characterization of the major Epstein-Barr virus-specific RNA in Burkitt lymphoma-derived cells. *J. Virol.* **41:**376–389.
3. **Baichwal, V. R., and B. Sugden.** 1988. Transformation of Balb 3T3 cells by the BNLF-1 gene of Epstein-Barr virus. *Oncogene* **2:**461–467.
4. **Baichwal, V. R., and B. Sugden.** 1989. The multiple membrane-spanning segments of the BNLF-1 oncogene from Epstein-Barr virus are required for transformation. *Oncogene* **4:**67–74.
5. **Birkenbach, M., D. Liebowitz, F. Wang, J. Sample, and E. Kieff.** 1989. Epstein-Barr virus latent infection membrane protein increases vimentin expression in human B-cell lines. *J. Virol.* **63:**4079–4084.
6. **Bodescot, M., O. Brison, and M. Perricaudet.** 1986. An Epstein-Barr virus transcription unit is at least 84 kilobases long. *Nucleic Acids Res.* **14:**2611–2620.
7. **Bodescot, M., and M. Perricaudet.** 1986. Epstein-Barr virus mRNAs produced by alternative splicing. *Nucleic Acids Res.* **14:**7103–7114.
8. **Bodescot, M., M. Perricaudet, and P. J. Farrell.** 1987. A promoter for the highly spliced EBNA family of RNAs of Epstein-Barr virus. *J. Virol.* **61:**3424–3430.
9. **Cordier, M., A. Calender, M. Billaud, U. Zimber, G. Rousslet, O. Pavlish, J. Banchereau, T. Tursz, G. Bornkamm, and G. M. Lenoir.** 1990. Stable transfection of Epstein-Barr virus (EBV) nuclear antigen 2 in lymphoma cells containing the EBV P3HR1 genome induces expression of B-cell activation molecules CD21 and CD23. *J. Virol.* **64:**1002–1013.
10. **Countryman, J., and G. Miller.** 1985. Activation of expression of latent Epstein-Barr herpesvirus after gene transfer with a small cloned subfragment of heterogeneous viral DNA. *Proc. Natl. Acad. Sci. USA* **82:**4085–4089.
11. **Dambaugh, T., F. Wang, K. Hennessy, E. Woodland, A. Rickinson, and E. Kieff.** 1986. Expression of the Epstein-Barr virus nuclear protein 2 in rodent cells. *J. Virol.* **59:**453–462.
12. **Desgranges, C., H. Wolf, G. de-Thé, K. Shanmugaratnam, N. Cammoun, R. Ellouz, G. Klein, K. Lennert, N. Muñoz, and H. zur Hausen.** 1975. Nasopharyngeal carcinoma. X. Presence of Epstein-Barr genomes in separated epithelial cells of tumours in patients from Singapore, Tunisia and Kenya. *Int. J. Cancer* **16:**7–15.
13. **Farrell, P. S.** 1989. Epstein-Barr virus genome. *Adv. Viral Oncol.* **8:**103–132.
14. **Fennewald, S., V. van Santen, and E. Kieff.** 1984. Nucleotide sequence of an mRNA tran-

scribed in latent growth-transforming virus infection indicates that it may encode a membrane protein. *J. Virol.* **51**:411–419.

15. **Fingeroth, J. D., J. J. Weis, T. F. Tedder, J. L. Strominger, P. A. Biro, and D. T. Fearon.** 1984. Epstein-Barr virus receptor of human B lymphocytes is the C3d receptor CR2. *Proc. Natl. Acad. Sci. USA* **81**:4510–4514.

16. **Hammerschmidt, W., and B. Sugden.** 1988. Identification and characterization of *oriLyt*, a lytic origin of DNA replication of Epstein-Barr virus. *Cell* **55**:427–433.

17. **Hammerschmidt, W., and B. Sugden.** 1989. Genetic analysis of immortalizing functions of Epstein-Barr virus in human B lymphocytes. *Nature* (London) **340**:393–397.

18. **Hammerschmidt, W., B. Sugden, and V. R. Baichwal.** 1989. The transforming domain alone of the latent membrane protein of Epstein-Barr virus is toxic to cells when expressed at high levels. *J. Virol.* **63**:2469–2475.

19. **Henderson, E., G. Miller, J. Robinson, and L. Heston.** 1977. Efficiency of transformation of lymphocytes by Epstein-Barr virus. *Virology* **76**:152–163.

20. **Hinuma, Y., M. Konn, J. Yamaguchi, D. J. Wudarski, J. R. Blakslee, Jr., and J. T. Grace, Jr.** 1967. Immunofluorescence and herpes-type virus particles in the P3HR-1 Burkitt lymphoma cell line. *J. Virol.* **1**:1045–1051.

21. **Howe, J. G., and M. D. Shu.** 1989. Epstein-Barr virus small RNA (EBER) genes: unique transcription units that combine RNA polymerase II and III promoter elements. *Cell* **57**:825–834.

22. **Jondal, M., and G. Klein.** 1973. Surface markers on human B and T lymphocytes. II. Presence of Epstein-Barr virus receptors on B lymphocytes. *J. Exp. Med.* **138**:1365–1378.

23. **Jones, C. H., S. D. Hayward, and D. R. Rawlins.** 1990. Interaction of lymphocyte-derived Epstein-Barr virus nuclear antigen EBNA-1 with its DNA-binding sites. *J. Virol.* **63**:101–110.

24. **Kieff, E., and D. Leibowitz.** 1990. Epstein-Barr virus and its replication, p. 1889–1920. *In* B. N. Fields and D. M. Knipe (ed.), *Virology*, 2nd ed. Raven Press, New York.

25. **Kintner, C., and B. Sugden.** 1981. Identification of antigenic determinants unique to the surfaces of cells transformed by Epstein-Barr virus. *Nature* (London) **294**:458–460.

26. **Klein, G.** 1989. Viral latency and transformation: the strategy of Epstein-Barr virus. *Cell* **58**:5–8.

27. **Klein, G., B. C. Giovanella, T. Lindahl, P. J. Fialkow, S. Singh, and J. S. Stehlin.** 1974. Direct evidence for the presence of Epstein-Barr virus DNA and nuclear antigen in malignant epithelial cells from patients with poorly differentiated carcinoma of the nasopharynx. *Proc. Natl. Acad. Sci. USA* **71**:4737–4741.

28. **Knutson, J. C.** 1990. The level of c-*fgr* RNA is increased by EBNA-2, an Epstein-Barr virus gene required for B-cell immortalization. *J. Virol.* **64**:2530–2536.

29. **Knutson, J. C., and B. Sugden.** 1989. Immortalization of B-lymphocytes by Epstein-Barr virus: what does the virus contribute to the cell? *Adv. Viral Oncol.* **8**:151–172.

30. **Lerner, M. R., N. C. Andrews, G. Miller, and J. A. Steitz.** 1981. Two small RNAs encoded by Epstein-Barr virus and complexed with proteins are precipitated by antibodies from patients with systemic lupus erythematosus. *Proc Natl. Acad. Sci. USA* **78**:805–809.

31. **Lupton, S., and A. J. Levine.** 1985. Mapping genetic elements of Epstein-Barr virus that facilitate extrachromosomal persistence of Epstein-Barr virus-derived plasmids in human cells. *Mol. Cell. Biol.* **5**:2533–2542.

32. **Manos, M. M., and Y. Gluzman.** 1984. Simian virus 40 large T-antigen point mutants that are defective in viral DNA replication but competent in oncogenic transformation. *Mol. Cell. Biol.* **4**:1125–1133.

33. **Mark, W., and B. Sugden.** 1982. Transformation of lymphocytes by Epstein-Barr virus requires only one-fourth of the viral genome. *Virology* **122**:431–443.

34. **Miller, G.** 1990. Epstein-Barr virus, p. 1921–1958. *In* B. N. Fields and D. M. Knipe (ed.), *Virology*, 2nd ed. Raven Press, New York.

35. **Miller, G., J. Robinson, L. Heston, and M. Lipman.** 1974. Differences between laboratory strains of Epstein-Barr virus based on immortalization, abortive infection, and interference. *Proc. Natl. Acad. Sci. USA* **71**:4006–4010.

36. **Pattengale, P. K., R. W. Smith, and P. Gerber.** 1973. Selective transformation of B lymphocytes by Epstein-Barr virus. *Lancet* **ii**:93–94.

37. **Rabson, M., L. Gradoville, L. Heston, and G. Miller.** 1982. Nonimmortalizing P3J-HR-1 Epstein-Barr virus: a deletion mutant of its transforming parent. *J. Virol.* **44**:834–844.

38. **Rawlins, D. R., G. Milman, S. D. Hayward, and G. S. Hayward.** 1985. Sequence-specific DNA binding of the Epstein-Barr virus nuclear antigen (EBNA-1) to clustered sites in the plasmid maintenance region. *Cell* **42**:859–868.

39. **Reedman, B. M., and G. Klein.** 1973. Cellular localization of an Epstein-Barr virus (EBV)-associated complement-fixing antigen in producer and nonproducer lymphoblastoid cell lines. *Int. J. Cancer* **11**:499–520.

40. **Reisman, D., and B. Sugden.** 1986. *trans* activation of an Epstein-Barr viral transcriptional enhancer by the Epstein-Barr viral nuclear antigen 1. *Mol. Cell. Biol.* **6**:3838–3846.

41. **Reisman, D., J. Yates, and B. Sugden.** 1985. A putative origin of replication of plasmids derived from Epstein-Barr virus is composed of two *cis*-acting components. *Mol. Cell. Biol.* **5**:410–413.

42. **Risser, R., and R. Pollack.** 1974. A non-selective analysis of SV40 transformation of mouse 3T3 cells. *Virology* **59**:477–491.

43. **Robinson, J., and D. Smith.** 1981. Kinetics of EBNA expression, cellular DNA synthesis, and mitosis. *Virology* **109**:336–343.

44. **Rosa, M. D., E. Gottlieb, M. R. Lerner, and J. A. Steitz.** 1981. Striking similarities are exhibited by two small Epstein-Barr virus-encoded ribonucleic acids and the adenovirus-associated ribonucleic acids VAI and VAII. *Mol. Cell. Biol.* **1**:785–796.

45. **Rowe, M., D. T. Rowe, C. D. Gregory, L. S. Young, P. J. Farrell, H. Rupani, and A. B. Rickinson.** 1987. Differences in B cell growth phenotype reflect novel patterns of Epstein-Barr virus latent gene expression in Burkitt's lymphoma cells. *EMBO J.* **6**:2743–2751.

46. **Shakespeare, W.** 1936. Hamlet, p. 1147–1193. *In* G. L. Kittredge (ed.), *Complete Works of Shakespeare*. Ginn and Co., Boston.

47. **Singer, B., and D. Grunberger.** 1983. *Molecular Biology of Mutagens and Carcinogens*, p. 33. Plenum Press, New York.

48. **Speck, S. H., and J. L. Strominger.** 1989. Transcription of Epstein-Barr virus in latently infected, growth-transformed lymphocytes. *Adv. Viral Oncol.* **8**:133–150.

49. **Stringer, J. R.** 1982. Mutant of simian virus 40 large T-antigen that is defective for viral DNA synthesis but competent for transformation of cultured rat cells. *J. Virol.* **42**:854–864.

50. **Sugden, B.** 1989. An intricate route to immortality. *Cell* **57**:5–7.

51. **Sugden, B., and W. Mark.** 1977. Clonal transformation of adult human leukocytes by Epstein-Barr virus. *J. Virol.* **23**:503–508.

52. **Sugden, B., M. Phelps, and J. Domoradzki.** 1979. Epstein-Barr virus DNA is amplified in transformed lymphocytes. *J. Virol.* **31**:590–595.

53. **Sugden, B., and N. Warren.** 1989. A promoter of Epstein-Barr virus that can function during latent infection can be transactivated by EBNA-1, a viral protein required for viral DNA replication during latent infection. *J. Virol.* **63**:2644–2649.

54. **Summers, W. P., E. A. Grogan, D. Shedd, M. Robert, C.-R. Liv, and G. Miller.** 1982. Stable expression in mouse cells of nuclear neoantigen after transfer of a 3.4-megadalton cloned fragment of Epstein-Barr virus DNA. *Proc. Natl. Acad. Sci. USA* **79**:5688–5692.

55. **Wang, D., D. Liebowitz, and E. Kieff.** 1985. An EBV membrane protein expressed in immortalized lymphocytes transforms established rodent cells. *Cell* **43**:831–840.

56. **Wang, D., D. Liebowitz, F. Wang, C. Gregory, A. Rickinson, R. Larson, T. Springer, and E.**

Kieff. 1988. Epstein-Barr virus latent infection membrane protein alters the human B-lymphocyte phenotype: deletion of the amino terminus abolishes activity. *J. Virol.* **62:**4173–4184.

57. **Wang, F., C. D. Gregory, M. Rowe, A. B. Rickinson, D. Wang, M. Birkenbach, H. Kikutani, T. Kishimoto, and E. Kieff.** 1987. Epstein-Barr virus nuclear antigen 2 specifically induces expression of the B-cell activation antigen CD23. *Proc. Natl. Acad. Sci. USA* **84:**3452–3456.

58. **Weber, P. C., M. D. Challberg, N. J. Nelson, J. Levine, and J. C. Glorioso.** 1988. Inversion events in the HSV-1 genome are directly mediated by the viral DNA replication machinery and lack sequence specificity. *Cell* **54:**369–381.

59. **Yates, J., N. Warren, D. Reisman, and B. Sugden.** 1984. A cis-acting element from the Epstein-Barr viral genome that permits stable replication of recombinant plasmids in latently infected cells. *Proc. Natl. Acad. Sci. USA* **81:**3806–3810.

60. **Yates, J. L., N. Warren, and B. Sugden.** 1985. Stable replication of plasmids derived from Epstein-Barr virus in various mammalian cells. *Nature* (London) **313:**812–815.

61. **zur Hausen, H., F. J. O'Neill, and U. K. Freese.** 1978. Persisting oncogenic herpesvirus induced by tumour promoter TPA. *Nature* (London) **272:**373–375.

Viruses That Affect the Immune System
Edited by Hung Y. Fan et al.
© 1991 American Society for Microbiology, Washington, DC 20005

Chapter 15

Epstein-Barr Virus Transcription in Latently Infected B Lymphocytes

Samuel H. Speck

HISTORICAL PERSPECTIVE

Epstein-Barr virus (EBV) is a human lymphotrophic herpesvirus which is ubiquitous among all human populations and is typically carried by more than 95% of the adult population. Once acquired, the virus persists in the host throughout life. It is the etiologic agent of infectious mononucleosis and is also closely associated with Burkitt's lymphoma and nasopharyngeal carcinoma. Consistent with its proposed role in oncogenesis, EBV infection of peripheral B lymphocytes in tissue culture results in a latent infection with a concomitant growth transformation of the infected lymphocytes (immortalization).

Since the discovery of EBV in the early 1960s, most of the research on this virus has been focused on understanding how it induces cellular proliferation. Early research was severely hampered by two major obstacles: (i) it was difficult to obtain large quantities of virus since no completely permissive cell culture system existed; and (ii) until very recently, it was not possible to generate mutant viruses. The latter point completely prevented the identification of viral genes essential for growth transformation. Thus, research on EBV relied on advances in recombinant DNA technology. By 1980 the EBV genome was cloned, and this led to mapping the viral genes expressed during the latent, growth-transforming life cycle of EBV.

Samuel H. Speck • Dana-Farber Cancer Institute, Harvard Medical School, Boston, Massachusetts 02115.

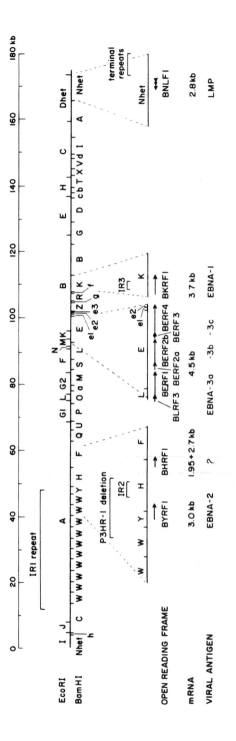

Early efforts employing solution hybridization kinetics to identify viral transcripts revealed that viral gene expression in latently infected lymphocytes was not localized to a specific area of the viral genome, but rather, many regions appeared to be active. With the advent of RNA blotting techniques, some of the viral transcripts were localized to specific regions of the viral genome. Furthermore, these studies revealed that many of the viral transcripts are surprisingly large (most greater than 3 kb) and that they are generally expressed at low abundance. Attempts to hybrid-select and in vitro translate these messages were unsuccessful, although this approach was successfully employed to map viral proteins associated with the lytic cycle. In retrospect, it is likely that failure to detect the viral antigens expressed during latency was due to (i) low abundance of the viral mRNAs and (ii) reliance on polyclonal human heterosera, which are now known to have low titers to most of the viral antigens expressed during latency.

Rapid progress in identifying viral proteins expressed during latency was made possible once the sequence of the viral genome was completed. To date, six viral nuclear antigens (EBNAs) and at least two and possibly three membrane antigens have been identified in latently infected lymphocytes. As indicated above, the genes which encode these antigens are not clustered to one region of the viral genome (Fig. 1). Indeed, the EBNAs are encoded by open reading frames (ORFs) which are spread out over nearly the entire left-hand two-thirds of the genome. This raises the question of how the virus regulates viral gene expression during latency while suppressing lytic gene expression. I am going to restrict my discussion of viral transcription to transcription of the EBNA genes, all of which are transcribed from the R strand.

While RNA blotting identified several regions of the viral genome that are transcriptionally active during latent infection, it could not account for the structures of these transcripts. This became apparent when the first viral nuclear antigen, EBNA-1, was identified. RNA blotting indicated that EBNA-1 is encoded by a single 3.7-kg transcript. However, the size of EBNA-1 suggested that only about 1.9 kb of coding information would be required. Transfection of a 2.9-kb fragment of the viral *Bam*HI-K genomic fragment was sufficient to obtain expression of an appropriate-sized EBNA-1 polypeptide. Furthermore, S1 nuclease analysis demonstrated that only a single 2-kb exon of the EBNA-1 transcript mapped to this region of the viral genome. This left unresolved the origin of the remaining 1.7 kb of the EBNA-1 tran-

---

Figure 1. *Bam*HI and *Eco*RI restriction endonuclease maps of the B95-8 strain of EBV. Regions of the viral genome that have been shown to be transcribed during viral latency are shown. The relevant ORFs and the viral proteins they encode are indicated. Note: the exons encoding EBNA-4 (see Fig. 2) and terminal protein (see Fig. 3) are not shown.

script. It became increasingly apparent that the structures of the viral transcripts in latently infected cells might be complex and that cloning these messages as cDNAs was necessary. (For a more complete review of the literature see references 6 and 14.)

IDENTIFICATION OF TWO VIRAL PROMOTERS INVOLVED IN DRIVING TRANSCRIPTION OF THE EBNA GENES

The first major clue into the structures of the transcripts encoding the EBNA genes came from cloning the EBNA-1 transcript. Initially great difficulty was encountered because of a G+C-rich repeat (IR3) within the EBNA-1 open reading frame, which presumably prevented efficient reverse transcription through this region. Thus, all the clones obtained were truncated and did not extend even to the 5' end of the EBNA-1 ORF. However, these cDNA clones did establish the utilization of the polyadenylation signal at the 3' end of the EBNA-1 ORF and that the unaccounted-for 1.7 kb of transcript mapped 5' to the EBNA-1 ORF. An additional problem, which is still not understood, is that the EBNA transcripts are severely underrepresented in all the cDNA libraries which we have screened.

To circumvent the problem of termination within the G+C-rich IR3 region, a primed cDNA library was constructed employing a specific oligonucleotide complementary to a region which lies approximately 250 nucleotides 3' to the putative EBNA-1 translation initiation codon and upstream of the G+C-rich IR3 region. One cDNA clone, JY-K2, was obtained which contained sequences upstream of the EBNA-1 ORF (13). The sequence of this clone revealed the presence of a number of mini-exons which are spread over nearly the entire left-hand half of the EBV genome, indicating that the primary transcript for EBNA-1 is greater than 70 kb in length. As startling as this result was, further interest was generated by the fact that several of the 5' exons in the JY-K2 cDNA were also present in another cDNA clone (Raji-T1) which was thought to reflect a portion of a viral mRNA present in latently infected lymphocytes (3). This was the first inkling that all the transcripts encoding the EBNAs share common 5' exons which are spliced to unique 3' coding domains.

The JY-K2 cDNA was not complete at the 5' end and therefore did not identify the viral promoter involved in driving transcription of the EBNA-1 gene. Subsequently a number of cDNA clones were characterized, and they confirmed the initial prediction that all the EBNA mRNAs share common 5' exons (see Fig. 2) (1–4, 10–13). All the transcripts characterized contained multiple copies of two small exons (W1 and W2 exons) encoded within the major internal repeat of EBV (3.1-kb direct repeats; *Bam*HI-W fragments).

In addition, most of these transcripts contained two or three small exons encoded within the adjacent BamHI-Y fragment (Y1, Y2, and Y3 exons). The data compiled from the analysis of cDNA clones indicated that the various mature EBNA messages are generated by alternative splicing from these common 5' exons to unique 3' coding exons.

While the analysis of the EBNA transcripts is far from complete, owing to the arduous task involved in obtaining these cDNA clones, two viral promoters involved in driving transcription of the EBNA genes have been identified. The 5' ends of three cDNA clones obtained from the latently infected lymphoblastoid cell line IB4 (IB4-WY1, IB4-T62, and IB4-W2.16) map just downstream of a consensus eucaryotic promoter contained within the major internal repeat of the virus (10–12). Both clones contain a single copy of a small exon, W0, which is spliced to a variable number of W1/W2 repeat exons. The promoter in BamHI-W (Wp) contains a TATA box just 30 bp upstream of the 5' end of the W0 exon, as well as an appropriately spaced CAAT box. A comprehensive study of the viral transcripts encoding the EBNAs in a marmoset lymphoblastoid cell line infected with EBV revealed a second promoter that mapped to the region of the BamHI-C fragment just upstream of the major internal repeat (2, 4, 5). Transcripts initiating from the BamHI-C promoter (Cp) contain two small exons, C1 and C2, which are spliced to a variable number of W1/W2 repeat exons.

ALTERNATIVE SPLICING DICTATES TRANSLATIONAL START IN EBNA TRANSCRIPTS

One of the findings from the characterization of EBNA cDNA clones was the identification of a previously unrecognized EBNA gene (EBNA-4) (2, 11, 12). The EBNA-4 gene is encoded largely by the small W1/W2 repeat exons from the major internal repeat of the virus. Since the ORF generated by the W1/W2 repeat exons remains in frame, this gives rise to a variable-sized, highly repetitive polypeptide which contains a unique carboxy terminus encoded by the Y1 and Y2 exons. The presence of the EBNA-4 coding exons in the 5' leader of all EBNA transcripts poses an interesting dilemma. Since downstream ORFs in polycistronic mRNAs are generally not efficiently translated in eukaryotic cells, this raises the question of whether there is a mechanism in EBV-infected lymphocytes for either (i) translating bicistronic mRNAs or (ii) skipping over the EBNA-4 ORF, thereby allowing efficient translation of the downstream EBNA gene.

Characterization of the cDNA clones containing an intact EBNA-4 gene revealed that the EBNA-4 translation initiation codon is generated by the splice junction between the 5' W0 exon and the first W1 exon (W1') (11, 12).

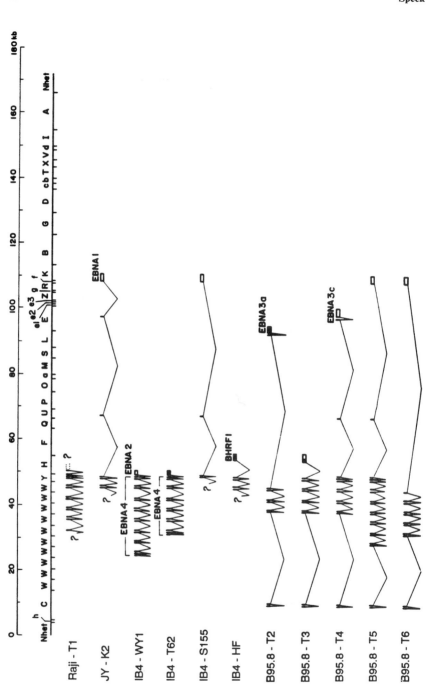

The W1′ exon is actually 5 nucleotides shorter than the other W1 exons, because the W0/W1′ splice junction utilizes a splice acceptor 5 nucleotides downstream of the splice acceptor utilized by the W2/W1 splice junctions. The last two bases of the W0 exon are AU while the first base of the W1′ exon is a G, which together generate the AUG translation initiation codon for EBNA-4. It should be emphasized that this is the only AUG codon in the entire EBNA-4 ORF.

If the W0 exon splices to the upstream splice acceptor for the W1 exon (as opposed to the acceptor of the W1′ exon), since the first base of the W1 exon is a C, the EBNA-4 translation initiation codon would not be generated. We assayed for the presence of W0/W1 and W0/W1′ splice junctions by employing splice junction-specific oligonucleotides (10). The results indicated that both splice junctions occur and that in fact the non-AUG splice junction is actually favored. We extended these studies to transcripts initiating from Cp, since the last two nucleotides of the C2 exon, like those of the W0 exon, are AU. As in the case with transcripts initiating from Wp, we found that both C2/W1 and C2/W1′ splice junctions exist. We have also isolated a cDNA clone from a cDNA library prepared from a latently infected cell line which contains the W0/W1 non-AUG splice junction (10). Thus, it appears that a mechanism exists in latently infected cells for generating EBNA transcripts which do not contain a translatable EBNA-4 gene.

MUTUALLY EXCLUSIVE PROMOTER UTILIZATION AND PROMOTER SWITCHING

The identification of two promoters involved in driving transcription of the EBNA genes raised the question of whether expression of a specific EBNA might be driven from only one of these promoters. However, a survey of a panel of Burkitt's lymphoma and lymphoblastoid cell lines revealed that in any given clonal cell line only one of these promoters is active, indicating that expression of all the EBNA genes can be driven by either Cp or Wp (15).

Figure 2. Schematic exon maps of representative cDNA clones representing rightward viral transcripts, shown with respect to the *Bam*HI restriction endonuclease map of the B95-8 strain of EBV. Filled boxes represent sequences present in the cDNA clone and open boxes represent proposed structures. The coding sequences for the known viral antigens present in latently infected lymphocytes are indicated. The cDNA clones are named according to the cell lines from which they were derived (Raji, a latently infected Burkitt's lymphoma cell line; JY and IB4, latently infected lymphoblastoid cell lines; B95.8, productively infected marmoset lymphoblastoid cell line). The exon structures from the indicated cDNA clones are from the following references: Raji-T1 (3); JY-K2 (13); IB4-WY1 (12); IB4-T62 and IB4-S155 (11); IB4-HF (1); B95.8-T2 (2); B95.8-T3, -T4, -T5 and -T6 (4).

There was no correlation between promoter usage and cellular phenotype (Burkitt's lymphoma or lymphoblastoid cells) or strain of infecting virus (type A or B virus).

To determine whether there is a fundamental difference in the transcription factors present in Cp- and Wp-utilizing cell lines, Cp and Wp activity of exogenous reporter constructs transfected into various Wp- and Cp-utilizing cell lines was determined (15, 16). The reporter constructs employed contained the region of the viral genome from the latency-associated origin of replication (oriP) to either the 5′ end of the W1 exon or the middle of the Y2 exon, and contained both Cp and Wp. These constructs were either linked to the bacterial gene encoding CAT for direct activity analyses or to the beta-globin gene for isolation of RNA and S1 nuclease protection assays. Analysis of promoter activity exhibited by an exogenous reporter construct in two different Cp-utilizing cell lines demonstrated Cp activity and no detectable Wp activity, consistent with the activity of the endogenous viral genomes. However, analysis of promoter activity exhibited by the exogenous reporter construct in two Wp-active cell lines revealed significant Cp activity and little or no Wp activity, in direct contrast to the resident viral genomes.

The results obtained by transfecting exogenous reporter constructs into Cp- or Wp-using cell lines demonstrated that the transcription factors necessary to drive Cp are present in Wp cell lines. This raises the question of whether Wp-using cell lines are Cp mutants. To begin to address this issue, we analyzed the viral DNA from three Wp-using cell lines by Southern blot analysis (16). This analysis revealed that the viral genomes from two of the cell lines, IB4 and X50-7, contained a 3.5-kb deletion in the BamHI-C fragment. Further analysis demonstrated that a short probe from the region of Cp does not hybridize to the BamHI-C fragment of the IB4 or X50-7 genomes, indicating that this region is deleted. These results support the hypothesis that Wp-utilizing cell lines harbor Cp mutant EBV genomes.

If indeed Cp is the promoter "normally" employed during established viral latency, then what is the role of Wp during viral infection? One possibility is that Wp is normally active during a phase of viral infection other than established latency. With regard to this hypothesis, one of the hallmarks of EBV infection of peripheral B lymphocytes is the partial differentiation of these infected cells to continuously proliferating lymphoblastoid cells. Thus, it is possible that initiation of viral gene expression in quiescent B lymphocytes might require a promoter specifically designed to function in that transcriptional milieu. If this is indeed the case, then the prediction is that Wp will be active during the initial stages of viral infection and that promoter activity will switch to Cp once the cells have differentiated to a defined stage.

To assess Cp and Wp activity during the initial stages of viral infection, peripheral B lymphocytes were infected and RNA was prepared at various time points postinfection (16). This revealed that Wp activity could be detected within the first 24 h of infection, while Cp activity in general could not be detected until nearly 6 days postinfection. This indicated that Wp is exclusively employed during the initiation of viral infection. It should be noted that Wp activity remained high throughout the first 6 days of infection, so there did not appear to be a concomitant down-regulation of Wp. At this time it is unclear why Wp activity persists in bulk culture infections. However, it may reflect an adaptive advantage to tissue culture conditions.

ORGANIZATION OF LATENT AND LYTIC VIRAL TRANSCRIPTION

While the discussion in this review has focused on the two promoters involved in driving transcription of the EBNA genes, there are two other viral promoters that are active during latency, namely, the promoters for the genes encoding the membrane proteins LMP (latent membrane protein) (7) and TP (terminal protein) (8). Interestingly, all four viral promoters which are active during latency are clustered around *ori*P (Fig. 3). Furthermore, *ori*P has been shown to contain an EBNA-1-dependent enhancer (9). Thus,

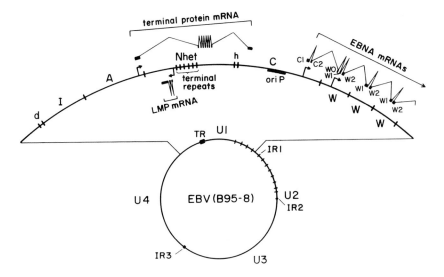

Figure 3. Organization of viral transcription in latently infected lymphocytes. The proximity to the latent origin of replication (*ori*P) of the viral promoters active during latency is shown. Only the left-hand-most Wp is illustrated.

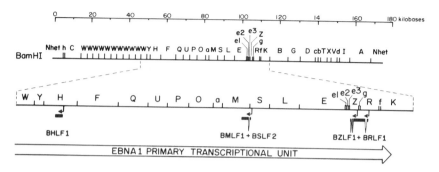

Figure 4. Schematic exon map of several viral genes involved in the induction of the viral lytic cycle. The genomic location and direction of transcription are shown with respect to the primary transcriptional unit of the viral gene encoding EBNA-1.

it is attractive to speculate that *ori*P is a central control region that is actively involved in regulating the expression of all these viral promoters.

A major question raised by the organization of latent transcription is why, since the control regions for all the viral genes expressed during latency are clustered together, are the EBNA genes spread over such a large area? One possible clue comes from the identification of the viral genes involved in the induction of the lytic cycle. A number of studies have shown that the BZLF1 gene product is capable of disrupting viral latency (reviewed in reference 14). In addition, the BRLF1 and BMLF1 genes are expressed early in the induction of the lytic cycle. All of these genes are positioned in the opposite orientation to the EBNA genes and map within the primary transcriptional unit of the EBNA 1 gene (Fig. 4). Thus, spreading out the regions encoding the various EBNA genes ensures continuous low-level transcription of the *R* strand through these regions of the genome, which may either prevent spurious transcription of the early lytic genes or quench low-level transcription of these genes by the presence of antisense RNA.

REFERENCES

1. **Austin, P. J., E. Flemington, D. N. Yandava, J. L. Strominger, and S. H. Speck.** 1988. Complex transcription of the Epstein-Barr virus BamHI H rightward open reading frame 1 (BHRF1) in latently and lytically infected B lymphocytes. *Proc. Natl. Acad. Sci. USA* **85**:3678–3682.

2. **Bodescot, M., O. Brison, and M. Perricaudet.** 1986. An Epstein-Barr virus transcription unit is at least 84 kilobases long. *Nucleic Acids Res.* **14**:2611–2620.

3. **Bodescot, M., B. Chambraud, P. Farrell, and M. Perricaudet.** 1984. Spliced RNA from the IR1-U2 region of Epstein-Barr virus: presence of an open reading frame for a repetitive polypeptide. *EMBO J.* **3**:1913–1917.

4. **Bodescot, M., and M. Perricaudet.** 1986. Epstein-Barr virus mRNAs produced by alternative splicing. *Nucleic Acids Res.* **14**:7103–7114.

5. **Bodescot, M., M. Perricaudet, and P. J. Farrell.** 1987. A promoter for the highly spliced EBNA family of RNAs of Epstein-Barr virus. *J. Virol.* **61**:3424–3430.

6. **Epstein, M. A., and B. G. Achong.** 1979. *The Epstein-Barr Virus.* Springer-Verlag, Berlin.

7. **Fennewald, S., V. van Santen, and E. Kieff.** 1984. Nucleotide sequence of an mRNA transcribed in latent growth-transforming virus infection indicates that it may encode a membrane protein. *J Virol.* **51**:411–419.

8. **Laux, G., M. Perricaudet, and P. Farrell.** 1988. A spliced Epstein-Barr virus gene expressed in immortalized lymphocytes is created by circularization of the linear genome. *EMBO J.* **7**:769–774.

9. **Reisman, D., and B. Sugden.** 1986. *trans*-Activation of an Epstein-Barr virus (EBV) transcriptional enhancer by the EBV nuclear antigen-1. *Mol. Cell. Biol.* **6**:3838–3846.

10. **Rogers, R. P., M. Woisetschlaeger, and S. H. Speck.** 1990. Alternative splicing dictates translational start in Epstein-Barr virus transcripts. *EMBO J.* **9**:2273–2277.

11. **Sample, J., M. Hummel, D. Braun, M. Birkenbach, and E. Kieff.** 1986. Nucleotide sequences of mRNAs encoding Epstein-Barr virus nuclear proteins: a probable transcriptional initiation site. *Proc. Natl. Acad. Sci. USA* **83**:5096–5100.

12. **Speck, S. H., A. J. Pfitzner, and J. L. Strominger.** 1986. An Epstein-Barr virus transcript from a latently infected, growth-transformed B cell line encodes a highly repetitive polypeptide. *Proc. Natl. Acad. Sci. USA* **83**:9298–9302.

13. **Speck, S. H., and J. L. Strominger.** 1985. Analysis of the transcript encoding the latent Epstein-Barr nuclear antigen 1: a potentially polycistronic message generated by long-range splicing of several exons. *Proc. Natl. Acad. Sci. USA* **82**:8305–8309.

14. **Speck, S. H., and J. L. Strominger.** 1989. Transcription of Epstein-Barr virus in latently infected, growth-transformed lymphocytes. *Adv. Viral Oncol.* **8**:133–150.

15. **Woisetschlaeger, M., J. L. Strominger, and S. H. Speck.** 1989. Mutually exclusive use of viral promoters in Epstein-Barr virus latently infected lymphocytes. *Proc. Natl. Acad. Sci. USA* **86**:6498–6504.

16. **Woisetschlaeger, M., C. N. Yandava, L. A. Furmanski, J. L. Strominger, and S. H. Speck.** 1990. Promoter switching in Epstein-Barr virus during the initial stages of infection of B lymphocytes. *Proc. Natl. Acad. Sci. USA* **87**:1725–1729.

Viruses That Affect the Immune System
Edited by Hung Y. Fan et al.
© 1991 American Society for Microbiology, Washington, DC 20005

Chapter 16

Expression of Cytomegalovirus in Mononuclear Cells

Jay A. Nelson, Carlos Ibanez, Rachel Schrier, Clayton Wiley, and Peter Ghazal

Cytomegalovirus (CMV) establishes a lifelong persistence in the host after primary infection, similar to other members of the human herpesvirus family. CMV is a notorious immunosuppressive agent. This property may contribute to the ability of the virus to persist in an individual by reducing the immunologic response of the host. Clinically, CMV immunosuppression is generally manifested in a depressed lymphocyte response to T-cell mitogens (3, 27, 44, 46), decreased natural killer (NK) cell activity (41, 44), and poor monocyte blastogenic response to mitogens such as concanavalin A or pokeweed mitogen antigen (PMA) (27, 41, 42). Fresh CMV isolates can abortively infect T lymphocytes in vitro, causing perturbation of function without morphologic alteration (34). However, the T cell may also be functionally impaired even without being productively infected (44). This phenomenon has been attributed to defects in monocyte function due to CMV infection. A variety of lymphoid cells are naturally infected by CMV, including polymorphonuclear leukocytes, monocytes, and T cells (19, 34, 44). Infection of these cells may not only produce the immunological defects elicited above, but also may act as a persistent source of the virus circulating through the body (43). Therefore, examining CMV replication in lymphoid cells will be important to understanding mechanisms of host cell interactions which result in immunosuppression.

Jay A. Nelson, Carlos Ibanez, and Peter Ghazal • Department of Immunology, Research Institute of Scripps Clinic, La Jolla, California 92037.   **Rachel Schrier and Clayton Wiley** • Department of Pathology, University of California at San Diego, La Jolla, California 92093.

HCMV REPLICATION IN PBMCs

Epidemiological studies have implicated transfused blood as an important source of CMV infection (1, 39, 62). Attempts to identify the blood cell types that are naturally infected by CMV have been hampered by the difficulty in detecting low levels of virus in a small percentage of cells. We have examined peripheral blood from 20 asymptomatic human CMV (HCMV)-seropositive individuals by in situ hybridization utilizing probes representing genes expressed during various stages of the viral replication cycle: immediate early (IE), early, and late (35, 43). We found that approximately 1 in 10^4 peripheral blood mononuclear cells (PBMCs) hybridize to the CMV probes. In these asymptomatic patients, hybridization was restricted to the IE probe, while signal was rarely detected with a late probe. Further experiments indicated that the primary infected populations are monocytes (up to 5%) and OKT4+ lymphocytes (up to 2.4%). A few OKT8+ cells (9.8%) but no NK or B cells demonstrated hybridization to the CMV probes. Culture of lymphocytes from CMV hybridization-positive donors on permissive human fibroblasts failed to produce infectious virus, consistent with previous observations. Therefore, CMV is present in a small number of circulating mononuclear cells in asymptomatic individuals, certain cells are preferentially infected (monocytes, OKT4 cells), and HCMV infection on mononuclear cells appears to be restricted.

This restriction of viral replication in monocytes and lymphocytes may be lifted by immunosuppression (i.e., immunosuppressive therapy during transplantation or acquired immune deficiency syndrome [AIDS]), allowing productive infection of these cells. Therefore, to determine whether the immune status of the individual affects viral replication in PBMCs, CMV expression was examined in the peripheral blood of AIDS patients. By in situ hybridization, individuals representing the three different clinical groups of AIDS (seropositive asymptomatic, AIDS-related complex, and overt AIDS) were all positive for the presence of CMV with approximately fivefold more cells hybridizing to the IE probe as compared to patients who were human immunodeficiency virus seronegative (35). However, the greatest contrast was observed with the use of the HCMV late probe, which hybridized with the same frequency as the IE probe. The results, taken together with the ease of isolating infectious virus from PBMCs of AIDS patients, suggest that CMV can productively infect these blood cells. The removal of restriction of virus expression in PBMCs may be due either to decreased immune surveillance or to a general activation of monocytes and T cells in AIDS patients. The latter scenario appears more likely since in vitro treatment of PBMCs with mitogens (phytohemagglutinin [PHA]) or antigens to which the donor is immune increases (10- to 20-fold) the number of cells hybridizing to CMV probes (unpublished observations). Therefore, the state of monocyte or T-cell activation may have profound effects on CMV replication.

Examination of other tissues during CMV disease processes indicates that mononuclear inflammatory cells are a major reservoir of CMV (15, 60). Biopsies of kidneys taken during the renal transplant process indicated that mononuclear cells were the predominant CMV-infected cell type. The mononuclear cells observed in the kidneys were part of a rejection process occurring in these patients. A high proportion of kidney-infiltrating cells were CMV positive in comparison to the frequency observed in the peripheral blood. From these studies, we hypothesized that CMV may infect more cells in the peripheral blood than can be detected by in situ hybridization. The process of mononuclear cell activation during the graft-versus-host response may also activate latent CMV infecting these cells.

Both the peripheral blood and kidney studies demonstrate that mononuclear cells are an important site of CMV infection and may act as vectors for dissemination of the virus throughout the body. Replication of the virus may be naturally restricted in undifferentiated cells. However, mitogenic or allogeneic stimulation of these cells to differentiate may induce cellular factors necessary for CMV replication.

HCMV EXPRESSION IS DEPENDENT ON THE STATE OF CELLULAR DIFFERENTIATION

The genes encoded by HCMV have been subdivided into three kinetic classes, immediate-early (IE), early, and late genes (7, 30, 57). The expression of the IE genes is dependent on host cell factors, while both early and late gene expression are interdependent on cellular and viral factors. The synthesis of IE proteins is required for the initiation of early gene expression since infection in the presence of inhibitors of protein synthesis prevents subsequent viral gene expression (7, 57). Studies to date have demonstrated that three distinct segments of the CMV genome are predominantly transcribed during IE expression (22, 25, 51, 58). Two regions of abundant IE gene expression, designated IE region 1 (IE-1) (51, 55) and IE region 2 (IE-2), originate from one of these segments (52). These genes are transcribed from a single promoter regulatory region located upstream of IE-1 (52), which has been referred to as the major IE promoter (MIEP) (2, 56). The levels of the major IE proteins synthesized in an infected cell are predicted to switch the viral genome from restrictive to extensive expression (35). Therefore, the transcription control of MIEP may describe a primary target for determining the state of the virus within a cell.

To examine regulation of CMV IE genes in undifferentiated and differentiated cells, we have utilized human teratocarcinoma cells (Tera-2) as a model system. CMV replicates in differentiated but not undifferentiated Tera-2 cells (16). By comparison of steady-state RNA levels and in vitro run-on transcription of nuclei, we demonstrated that the major IE gene is inactive in undifferentiated but active in differentiated Tera-2 cells. Thus, the block in HCMV replication in these cells is at the transcriptional level of the MIEP (36).

Analysis of chromatin characteristics of active and inactive genes has identified sites hypersensitive to nuclease attack which correlate with important *cis*-acting regulatory elements. A comparison of the structural features of chromatin on the promoter regulatory region with the active and inactive IE genes demonstrated the presence of constitutive and inducible DNase I-hypersensitive sites. The majority of the constitutive sites existed between −175 and −525 relative to the start of transcription (Fig. 1) (36). This region was shown to have simian virus 40 enhancer function (2). In contrast, the inducible DNase I-hypersensitive sites were located outside this region between −650 and −975 as well as in an area within the first exon. Since this is one of the primary viral genes to be turned on when the virus enters the cell, the increased nuclease hypersensitivity in the active gene may reflect an altered chromatin conformation which is favorable for the binding of transcription factors. Furthermore, altered DNA conformations may also be involved. Altered DNA structures were detected in sequences distal to the enhancer during the active transcription of these genes using a chemical (carcinogen chloracetaldehyde) as a probe (24).

To test the significance of the upstream inducible DNase I-hypersensitive

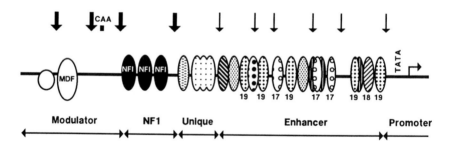

Figure 1. Diagram of the structure of the MIEP regulatory region. The line represents the 5'-flanking sequences with the blobs marking regions of in vitro DNase I protection. The different coded blobs positioned over these protected areas illustrate the distinctive nature and multiplicity of interacting nuclear proteins. The numbers refer to the repeat elements protected. Slim vertical arrows mark in vivo constitutive nuclease-hypersensitive sites, while the bold vertical arrows mark inducible hypersensitive sites. CAA marks the interaction point of the carcinogen chloroacetylaldehyde. Abbreviations: NF1, nuclear factor 1 protein; MDF, modulator dyad factor.

sites, reporter constructs under the control of variously deleted portions of MIEP were assayed in transient transfection experiments using differentiated and undifferentiated Tera-2 cells (37). These experiments identified a novel regulatory region between −750 and −1145. Deletion of this region causes a 10-fold increase in MIEP activity in undifferentiated Tera-2 cells (37). However, in permissive differentiated Tera-2 cells, removal of this regulatory region results in a decrease in activity. Therefore, the MIEP is a complex element which contains a dual-function *cis* regulatory element between −750 and −1145 which negatively modulates expression in undifferentiated cells but positively influences expression in differentiated cells. On the basis of these observations we have designated this region as the "modulator."

Understanding the mechanism of CMV activation in Tera-2 cells will be important and may mimic regulation of MIEP in primary monocyte/macrophages. We have recently developed a human monocyte system in which expression of CMV MIEP is highly dependent on the state of cellular differentiation. Preliminary results suggest that the CMV expression in primary monocyte/macrophages is dependent on activated T-cell contact with the monocytes immediately upon plating. Monocyte/macrophages in this system undergo differentiation into multinucleated giant cells (Fig. 2) and can be maintained in culture up to 40 days postdifferentiation. By transient transfection experiments we have found that expression of the CMV MIEP in the undifferentiated monocytes is restricted whereas differentiation removes the block similar to the Tera-2 system (unpublished observations).

Therefore, by utilizing these in vitro cell culture systems we suggest that the state of the virus within the cell is dependent on the level of IE gene expression. In addition, the corresponding regulatory *cis*-acting sequences and their respective *trans*-acting factors are functionally dependent on the differentiation state of the cell.

REGULATION OF THE CMV MIEP IN LYMPHOID CELLS

Expression of CMV genes from the major IE region is vitally important for productive infection of a cell. The MIEP which controls these genes is highly dependent on cellular transcription factors for activity. Therefore, identification of *cis* elements and cellular transcription factors interacting with these sequences will be critically important in understanding tissue-specific expression of this virus. The MIEP regulatory region is composed of distinct control domains which include an RNA polymerase II promoter between +1 and −50 (12), a strong enhancer between −50 and −530 (2, 11, 56), a unique sequence region (between −530 and −640) (14) adjacent to a cluster of NF-1/CTF binding sites between −650 and −750 (17), and the modulator se-

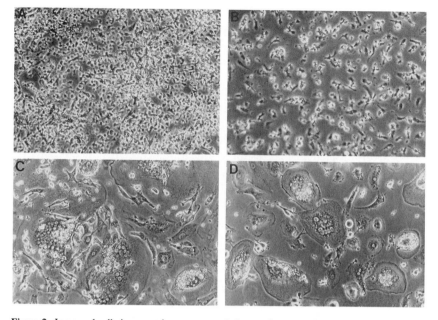

Figure 2. Increased cell size, protein content, and changes in morphology suggest that cultured monocytes stimulated with concanavalin A and nonadherent lymphocytes mature into multinucleated macrophages. Cultures of monocytes at various stages of differentiation: (A) 1 day; (B) 6 days; (C) 12 days; and (D) 25 days. Magnification, ×200. Preliminary results of activation assays, which include levels of 5′ nucleotidase, β-glucuronidase, and superoxide activities, are consistent with other reports describing the differentiation process.

quence between −750 and −1145 (37). In this section we will describe the activity of the modulator in lymphoid cells and discuss those inducible elements within the enhancer which respond to stimulation by cyclic AMP, PHA, and PMA.

Modulator

The modulator region as described above negatively affects MIEP function in undifferentiated nonpermissive Tera-2 cells while positively influencing transcription from MIEP in the differentiated cells. We have examined whether the modulator sequences confer cell specificity for expression in other types of cells (28). CAT reporter constructs containing the MIEP with and without the modulator were assayed in several different cell types (epithelial, T cells, B cells). Either the constructs were transiently transfected and analyzed, or transcriptionally active extracts were prepared and assayed. The effect of the modulator sequence on CAT activity in the transient transfection experiments or levels of transcription from MIEP in the in vitro transcription assays differed depending on the cell type. A negative effect was observed in H-9 (T-cell), CEM (T-cell), and SW480 (epithelial) cells, but expression in Jurkat (T-cell), 293 (kidney), Raji (B-cell), Namalwa (B-cell), and U937 (monocyte/macrophage) cells was unaffected by the presence of the modulator sequence. These results indicate that the HCMV modulator sequence can influence MIEP activity in cell types other than teratocarcinoma cells. In particular, the modulator negatively affects expression in cell types (T cells and epithelial cells) that are important during natural infection.

To determine whether protein-DNA complexes in the modulator region correlated with transcriptional activity, the nuclear extracts described above were utilized in mobility shift assays. Similar migrating nuclear protein-DNA complexes which formed between the modulator region and the various nuclear extracts were detected. A simple correlation was not observed between modulator activity and a specific migrating nucleoprotein complex (28). Nucleoproteins from these extracts were found to interact with distinct regions of the modulator. A major complex (complex C in reference 28) mapped to sequences containing a large dyad symmetry by the mobility shift assay with different and overlapping restriction fragments. The protein(s) binding to this sequence has been designated the modulator dyad factors(s) (MDF). Similar observations were made by Shelburn et al. (45) in the Tera-2 system. These investigators found that undifferentiated cells contain a specific nuclear factor that binds to this dyad symmetry. A major decrease in this factor was observed upon differentiation of the Tera-2 cells. They also found that deletion of this sequence resulted in increased expression in undifferentiated nonpermissive cells. They hypothesize that this novel factor is a good candidate for a differentiation-specific negative regulatory factor for IE expression. The above studies demonstrate that the modulator sequence exerts differential activities in a variety of cells. We suggest that the unequal pattern of activity of the modulator sequence may be an important determinant toward viral latency and reactivation.

Enhancer

A remarkable array of short repetitive sequences ranging in size from 16–17 nucleotides to 21 nucleotides occur between −510 and −50 (2, 11, 56). This region is an important regulatory region (51) and possesses strong enhancer activity (2). We have been able to reproduce partially the activity of the MIEP enhancer in vitro using cell-free nuclear extracts (13). A competition assay in vitro demonstrated that the stimulation of transcription by the enhancer sequences was due to specific *trans*-acting factors (13). In vitro DNase I footprinting demonstrated the existence of multiple distinct cellular proteins that bind specifically to the various repeat sequences as well as to unique sequences (11, 13). A summary of the sequences contacted by sequence-specific nuclear proteins is depicted in Fig. 3. In vitro transcription oligo-nucleotide-competition assays (13) clearly demonstrated the requirement for the repeat elements as well as unique sequence binding sites to provide enhancer function. Importantly, this work suggests that the enhancer requires the coordinated action of several distinct elements, some of which are reiterated while others only occur once. The MIEP enhancer can be induced above its strong constitutive levels by cyclic AMP, PHA, and PMA and is particularly inducible by these agents in lymphoid cells (4, 5, 21, 48). The transduction pathways for these signals in lymphoid cells mimic events that lead to the activation of the cells of the immune system. The 19-bp repeat element described above mediates the response to cyclic AMP and PHA (4, 21, 48), suggesting that the proteins binding this element belong to the ATF/ CREB family of proteins (32, 33). The 18-bp repeat element is responsive to PMA and PHA stimulation and has been suggested to bind NF-κB (5). However, this motif also binds other ubiquitous proteins, and it has yet to be determined whether or not NF-κB binds to the MIEP 18-bp element.

Summing our current knowledge of the MIEP region, we conclude that the 5'-flanking sequence of the MIEP is highly complex and appears to be composed of several domains. These domains interface with a wide variety of cellular transcription factors. The known regulatory domains and protein-DNA interactions of the HCMV MIEP are summarized in Fig. 1. The elements appear to be acting coordinately to contribute to chromatin conformations important for accessibility to these transcription factors and subsequent gene expression. We think that the overall activity of the negative and positive domains of the HCMV MIEP within any given cell determines the final level of IE expression.

COORDINATE CONTROL OF HCMV EXPRESSION BY IE PROTEINS

As stated above, the MIEP regulates transcription of several different transcripts that have been divided into two regions designated IE regions 1 and 2 (IE-1 and IE-2) (Fig. 4) (51, 52, 55). These two regions code for a family of mRNAs that differ due to splicing and which code for a series of unique and related proteins (49, 52). The IE-1 codes for the major IE protein (72 kDa) which originates from a 1,950-nucleotide mRNA (51). The IE-2 codes for several proteins, some of which share sequences with IE-1 (49, 50, 52). The predominant IE proteins from region 2 are the 86-kDa and the 55-kDa proteins. The 55-kDa protein is expressed only under IE conditions, while

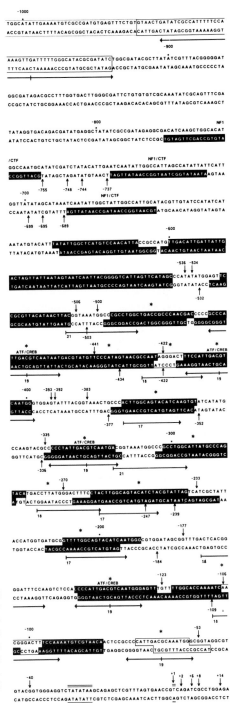

Figure 3. Sites of protein-DNA interaction on the MIEP regulatory region. Numbers represent the distance in base pairs located upstream of the transcription start site (+1). Sequences protected from DNase I cleavage are marked by reverse printing (strong protection) or a box (weak protection). Vertical arrows mark positions of enhanced protein-induced DNase I cleavage in vitro. Repeated sequence elements are designated by horizontal arrows with the respective repeat size indicated below. Those binding sites shown to be functional are marked with an asterisk.

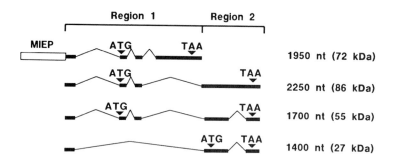

Figure 4. HCMV IE transcription unit in regions 1 and 2 showing the splicing pattern and IE proteins encoded by these mRNAs. Black boxes mark the exons while the thin interconnecting lines represent intronic sequences. ATG and TAA show the location of the initiation and termination codons, respectively.

the 86-kDa protein persists throughout infection. At late times, a 40-kDa protein is expressed from region 2 (49). Numerous studies (6, 9, 18, 54) have shown that viral protein products from these regions are capable of *trans*-activating heterologous promoters. Recently, proteins from IE-1 and -2 were shown to exhibit promoter-specific differences in their modes of action by either transactivating or repressing early and MIE promoters (38, 50). In transient cotransfection experiments the separate 86-kDa and 72-kDa proteins were found to minimally transactivate the CMV early promoter. However, the synergistic interaction of the two proteins resulted in optimal promoter activity. In contrast, the greatest MIEP activity occurred in the presence of the 72-kDa protein, whereas expression was specifically repressed with 86-kDa protein. Synergistically, the 72-kDa and 86-kDa proteins increased MIEP activity relative to the 86-kDa protein alone but repressed activity relative to the 72-kDa. A model for IE protein function is shown in Fig. 5. The studies show that the 72-kDa and 86-kDa proteins interact to regulate subsequent CMV gene expression. While the 86-kDa protein is involved in both activation and repression, its expression is regulated by the presence of the 72-kDa protein. Therefore, the 72-kDa protein produced in the infected cell may be the determining factor in the commitment to activation or repression.

Examination of the primary sequence structure of the 72-kDa, 55-kDa, and 86-kDa proteins reveals several interesting features characteristic of nuclear proteins and transcription factors (Fig. 6). By insertional mutagenesis and expression of the cDNAs encoding these products, the protein domains involved in regulation were investigated (50). Results from transfection experiments with both MIEP and a CMV early promoter identified three key regions as important for activation or repression. The first domain resides within exon 3 between amino acids 32 and 59 and is required for activation

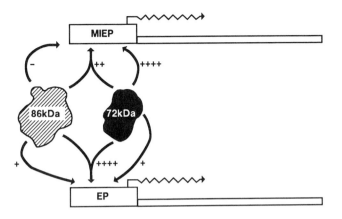

Figure 5. Model of IE protein function. Influence of the IE 72- and 86-kDa proteins on the MIEP and the early polymerase promoter (EP) is depicted. Direction of transcription from the promoters is indicated by the offset arrows. Curved arrows from the proteins indicate interaction of the proteins and their subsequent influence on the promoters. The relative levels of activation (+) and repression (−) of the promoters by IE proteins are indicated.

of the early promoter. This region is leucine rich, with a general motif of leucine-X_3-leucine. The importance of these leucine residues is unknown. A proline-rich region between amino acids 11 and 25 exists at the amino terminus. A proline-rich domain was recently identified as the transcription factor CTF/NF-1, corresponding to a transcriptional activation domain (31).

A second potential domain unique for the 72-kDa protein resides in exon 4. Examination of the primary sequence of this exon reveals the presence of a single putative zinc finger (20, 31, 40) located between amino acids 266 and 285, adjacent to a putative leucine zipper motif (26) between amino acids 294 and 325. The leucine zipper appears to promote protein-protein interactions to effect DNA binding. The functionality and importance of the zinc finger and leucine zipper motifs must await further experimentation. Analysis of the amino terminus of exon 4 reveals the presence of long segments of polyglutamic acid residues which may form potential acidic activation domains. In addition, exon 4 also contains polyserine stretches which may be potential sites of phosphorylation (10, 23, 59, 61).

A third domain in IE-2 exists between amino acids 359 and 540, encompassing a putative zinc finger distinct from the motif in the 72-kDa protein as well as a leucine-rich domain which weakly conforms to the leucine zipper motif. This region acts as both an activator of the early promoter and a repressor of the MIEP. Whether or not the activator repressor domains are identical is not known; however, two mutants which define this region are

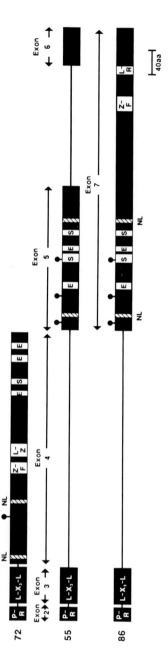
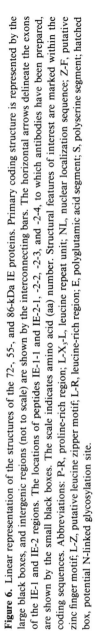

Figure 6. Linear representation of the structures of the 72-, 55-, and 86-kDa IE proteins. Primary coding structure is represented by the large black boxes, and intergenic regions (not to scale) are shown by the interconnecting bars. The horizontal arrows delineate the exons of the IE-1 and IE-2 regions. The locations of peptides IE-1-1 and IE-2-1, -2-2, -2-3, and -2-4, to which antibodies have been prepared, are shown by the small black boxes. The scale indicates amino acid (aa) number. Structural features of interest are marked within the coding sequences. Abbreviations: P-R, proline-rich region; L-X₃-L, leucine repeat unit; NL, nuclear localization sequence; Z-F, putative zinc finger motif; L-Z, putative leucine zipper motif; L-R, leucine-rich region; E, polyglutamic acid segment; S, polyserine segment; hatched box, potential N-linked glycosylation site.

phenotypically identical. Therefore, common and unique protein domains within the IE proteins mediate activation and repression.

Other amino acid motifs present in the IE proteins include potential nuclear localization signals which are characterized by the clustering of four to six basic amino acids (8) as well as N-linked glycosylation sites (29). The locations of the putative regulatory domains in the IE proteins are summarized in Fig. 6.

CONCLUSION

We have described the importance of the peripheral blood mononuclear cells and their state of differentiation in the biology of HCMV. Model tissue culture systems that were utilized to study how the differentiation state may regulate viral gene expression strongly indicate the importance of cellular elements influencing transcription of IE genes and the subsequent viral state. The interactions of cellular transcription factors with the MIEP region and expression of the IE products may constitute the earliest event directing the viral replication cycle. Therefore, identification of *cis* and *trans* regulatory factors controlling the expression of these IE genes is expected to be pivotal in determining the mechanisms of latency and reactivation.

Acknowledgments. This is publication no. 6292-IMM from the Research Institute of Scripps Clinic. This work was supported by grants AI-21640, CA-50151, and AG-04342. J.A.N. is a recipient of a Faculty Award from the American Cancer Society. We thank Margaret Stone for the excellent preparation and typing of this chapter.

REFERENCES

1. **Ballard, R. A., W. L. Drew, K. G. Hufnagle, and P. A. Riedel.** 1979. Acquired cytomegalovirus infection in preterm infants. *Am. J. Dis. Child.* **133:**482–485.
2. **Boshart, M., F. Weber, G. Jahn, K. Dorsch-Hasler, B. Fleckenstein, and W. Schaffner.** 1985. A very strong enhancer is located upstream of an immediate-early gene of human cytomegalovirus. *Cell* **41:**521–530.
3. **Carney, W. P., R. H. Rubin, R. A. Hoffman, W. P. Hansen, K. Healey, and M. S. Hirsch.** 1981. Analysis of T lymphocyte subsets in cytomegalovirus mononucleosis. *J. Immunol.* **126:**2114–2116.
4. **Chang, Y.-N., S. Crawford, J. Stall, D. R. Rawlins, K. T. Jeang, and G. S. Hayward.** 1990. The palindromic series 1 repeats in the simian cytomegalovirus major immediate-early promoter behave as both strong basal enhancers and cyclic AMP response elements. *J. Virol.* **64:**264–277.
5. **Cherrington, J. M., and E. S. Mocarski.** 1989. Human cytomegalovirus ie1 transactivates the α promoter-enhancer via an 18-base-pair repeat element. *J. Virol.* **63:**1435–1440.
6. **Davis, M. G., S. C. Kenney, J. Kamine, J. S. Pagano, and E.-S. Huang.** 1987. Immediate-

early gene region of human cytomegalovirus trans-activates the promoter of human immunodeficiency virus. *Proc. Natl. Acad. Sci. USA* **84**:8642–8646.

7. **DeMarchi, J. M.** 1981. Human cytomegalovirus DNA: restriction enzyme cleavage maps and map locations for immediate-early, early, and late RNAs. *Virology* **114**:23–38.

8. **Dingwall, C., and R. A. Laskey.** 1986. Protein import into the cell nucleus. *Annu. Rev. Cell Biol.* **2**:367–390.

9. **Everett, R. D.** 1984. Trans-activation of transcription of herpes virus products: requirements for two HSV-1 immediate early polypeptides for maximum activity. *EMBO J.* **3**:3145–3151.

10. **Furst, P., S. Hu, R. Hackett, and D. Hamer.** 1988. Copper activates metallothionein gene transcription by altering the conformation of a specific DNA binding protein. *Cell* **55**:705–717.

11. **Ghazal, P., H. Lubon, B. Fleckenstein, and L. Hennighausen.** 1987. Binding of transcription factors and creation of a large nucleoprotein complex on the human cytomegalovirus enhancer. *Proc. Natl. Acad. Sci. USA* **84**:3658–3662.

12. **Ghazal, P., H. Lubon, and L. Hennighausen.** 1988. Specific interactions between transcription factors and the promoter-regulatory region of the human cytomegalovirus major immediate-early gene. *J. Virol.* **62**:1076–1079.

13. **Ghazal, P., H. Lubon, and L. Hennighausen.** 1988. Multiple sequence-specific transcription factors modulate cytomegalovirus enhancer activity in vitro. *Mol. Cell. Biol.* **8**:1809–1811.

14. **Ghazal, P., H. Lubon, C. Reynolds-Kohler, L. Hennighausen, and J. A. Nelson.** 1990. Interactions between cellular regulatory proteins and a unique sequence region in the human cytomegalovirus major immediate-early promoter. *Virology* **174**:18–25.

15. **Gnann, J. W., J. Ahlmen, C. Svalander, L. Olding, M. B. A. Oldstone, and J. A. Nelson.** 1988. Inflammatory cells in transplanted kidneys are infected by human cytomegalovirus. *Am. J. Pathol.* **132**:239–248.

16. **Gonczol, E., P. W. Andrews, and S. A. Plotkin.** 1984. Cytomegalovirus replicates in differentiated but not undifferentiated human embryonal carcinoma cells. *Science* **224**:159–161.

17. **Hennighausen, L., and B. Fleckenstein.** 1986. Nuclear factor 1 interacts with five DNA elements in the promoter region of the human cytomegalovirus major immediate-early gene. *EMBO J.* **5**:1367–1371.

18. **Hermiston, T., C. Malone, P. Witte, and M. Stinski.** 1987. Identification and characterization of the human cytomegalovirus immediate-early region 2 gene that stimulates gene expression from an inducible promoter. *J. Virol.* **61**:3214–3221.

19. **Ho, M.** 1982. *Cytomegalovirus Biology and Infection*, p. 119–129. Plenum Medical Book Co., New York.

20. **Huckaby, C. S., O. M. Conneely, W. G. Beattie, A. W. Dabron, M. J. Tsai, and B. W. O'Malley.** 1987. Structure of the chromosomal chicken progesterone receptor gene. *Proc. Natl. Acad. Sci. USA* **84**:8380–8384.

21. **Hunninghake, G. W., M. M. Monick, B. Lin, and M. Stinski.** 1989. The promoter-regulatory region of the major immediate-early gene of human cytomegalovirus responds to T-lymphocyte stimulation and contains functional cyclic AMP-response elements. *J. Virol.* **63**:3026–3033.

22. **Jahn, G., E. Knust, H. Schmolla, T. Sarre, J. A. Nelson, J. K. McDougall, and B. Fleckenstein.** 1984. Predominant immediate-early transcripts of human cytomegalovirus AD169. *J. Virol.* **49**:363–370.

23. **Kadonaga, J. T., K. R. Cavner, F. R. Masiarz, and R. Tjian.** 1987. Isolation of cDNA encoding transcription factor Sp1 and functional analysis of the DNA binding domain. *Cell* **51**:1079–1090.

24. **Kohwi-Shigematsu, T., and J. A. Nelson.** 1988. The chemical carcinogen, chloracetaldehyde, modifies a specific site within the regulatory sequence of human cytomegalovirus major immediate gene in vivo. *Mol. Carcinog.* **1**:20–25.

25. Kouzarides, T., A. T. Bankier, S. C. Satchwell, E. Preddy, and B. G. Barrell. 1988. An immediate-early gene of human cytomegalovirus encodes a potential membrane glycoprotein. *Virology* **165**:151–164.

26. Landschulz, W. H., P. F. Johnson, and S. L. McKnight. 1988. The leucine zipper: a hypothetical structure common to a new class of DNA binding proteins. *Science* **240**:1759–1764.

27. Levin, M. J., C. R. Rinaldo, P. L. Leavy, J. A. Zaia, and M. S. Hirsch. 1979. Immune response to herpes virus antigens in adults with acute cytomegalovirus mononucleosis. *J. Infect. Dis.* **140**:851–857.

28. Lubon, H., P. Ghazal, L. Hennighausen, C. Reynolds-Kohler, C. Lockshin, and J. A. Nelson. 1989. Cell-specific activity of the modulator region in the human cytomegalovirus major immediate-early gene. *Mol. Cell. Biol.* **9**:1342–1345.

29. Marshall, R. D. 1972. Glycoproteins. *Annu. Rev. Biochem.* **41**:673–701.

30. McDonough, S., and D. Spector. 1985. Transcription in human fibroblasts permissively infected by human cytomegalovirus strain AD169. *Virology* **125**:31–46.

31. Mermod, N., E. A. O'Neill, T. J. Kelly, and R. Tjian. 1989. The proline-rich transcriptional activator CTF/NF1 is distinct from the replication and DNA binding domain. *Cell* **58**:741–753.

32. Montminy, M. R., and L. M. Bilezsikjian. 1987. Binding of a nuclear protein to the cyclic AMP response element of the somatostatin gene. *Nature* (London) **328**:175–178.

33. Montminy, M. R., K. A. Sevarino, J. A. Wagner, G. Mandel, and R. H. Goodman. 1986. Identification of a cyclic-AMP-responsive element within the rat somatostatin gene. *Proc. Natl. Acad. Sci. USA* **83**:6682–6686.

34. Myerson, D., R. C. Hackman, J. A. Nelson, D. C. Ward, and J. K. McDougall. 1984. Widespread presence of histologically occult cytomegalovirus. *Hum. Pathol.* **15**:430–439.

35. Nelson, J. A., J. W. Gnann, and P. Ghazal. 1990. Regulation and tissue specific expression of human cytomegalovirus. *Curr. Top. Microbiol. Immunol.* **154**:77–103.

36. Nelson, J. A., and M. Groudine. 1986. Transcriptional regulation of the human cytomegalovirus major immediate-early gene is associated with induction of DNase I hyper-sensitive sites. *Mol. Cell. Biol.* **6**:452–461.

37. Nelson, J. A., C. Reynolds-Kohler, and B. Smith. 1987. Negative and positive regulation by a short segment in the 5'-flanking region of the human cytomegalovirus major immediate-early gene. *Mol. Cell. Biol.* **7**:4125–4129.

38. Pizzorno, M. C., P. O'Hare, L. Sha, R. L. La Femina, and G. S. Hayward. 1988. Transactivation and autoregulation of gene expression by the immediate-early region 2 gene products of human cytomegalovirus. *J. Virol.* **62**:1167–1179.

39. Prince, A. M., W. Szmuness, S. J. Millian, and D. S. David. 1971. A serologic study of cytomegalovirus infections associated with blood transfusions. *N. Engl. J. Med.* **284**:1125–1131.

40. Rhodes, D., and A. Klug. 1988. "Zinc fingers": a novel motif for nucleic acid binding. *Nucleic Acids Mol. Biol.* **2**:149–166.

41. Rinaldo, C. R., Jr., P. H. Black, and M. S. Hirsch. 1977. Virus-leukocyte interactions in cytomegalovirus mononucleosis. *J. Infect. Dis.* **136**:667–678.

42. Rinaldo, C. R., Jr., W. P. Carney, B. S. Richter, P. H. Black, and M. S. Hirsch. 1980. Mechanisms of immunosuppression in cytomegalovirus mononucleosis. *J. Infect. Dis.* **141**:488–495.

43. Schrier, R. D., J. A. Nelson, and M. B. A. Oldstone. 1985. Detection of human cytomegalovirus in peripheral blood lymphocytes in a natural infection. *Science* **230**:1048–1051.

44. Schrier, R. D., G. Rice, and M. B. A. Oldstone. 1986. Suppression of NK activity and T cell proliferation induced by fresh isolates of cytomegalovirus. *J. Infect. Dis.* **153**:1084–1091.

45. Shelburn, S. L., S. K. Kothari, J. G. P. Sissons, and J. H. Sinclair. 1989. Repression of

human cytomegalovirus gene expression associated with a novel immediate early regulatory region binding factor. *Nucleic Acids Res.* **17**:9165–9171.

46. **Sissons, J. G. P.** 1986. The immunology of cytomegalovirus infection. *J. R. Coll. Physicians, Lond.* **20**:40–44.

47. **Southern, P., and M. B. A. Oldstone.** 1986. Medical consequences of persistent viral infection. *N. Engl. J. Med.* **314**:359–367.

48. **Stamminger, T., H. Fickenscher, and B. Fleckenstein.** 1990. Cell type-specific induction of the major immediate early enhancer of human cytomegalovirus by cyclic AMP. *J. Gen. Virol.* **71**:105–113.

49. **Stenberg, R. M., A. S. Depto, J. Fortney, and J. A. Nelson.** 1989. Regulated expression of early and late RNAs and proteins from human cytomegalovirus immediate-early gene region. *J. Virol.* **63**:2699–2708.

50. **Stenberg, R. M., J. Fortney, S. W. Barlow, B. P. Magrane, J. A. Nelson, and P. Ghazal.** 1990. Promoter-specific *trans* activation and repression by human cytomegalovirus immediate-early proteins involves common and unique protein domains. *J. Virol.* **64**:1556–1565.

51. **Stenberg, R. M., D. R. Thomsen, and M. F. Stinski.** 1984. Structural analysis of the major immediate-early gene of human cytomegalovirus. *J. Virol.* **49**:190–199.

52. **Stenberg, R. M., P. R. Witte, and M. F. Stinski.** 1985. Multiple spliced and unspliced transcripts from human cytomegalovirus immediate-early region 2 and evidence for a common initiation site within immediate-early region 1. *J. Virol.* **56**:665–675.

53. **Stinski, M. F.** 1983. The molecular biology of cytomegaloviruses, p. 67–113. *In* B. Roizman (ed.), *The Herpes Viruses*, vol. 2. Plenum Publishing Corp., New York.

54. **Stinski, M. F., and T. S. Roehr.** 1985. Activation of the major immediate-early gene of human cytomegalovirus by *cis*-acting elements in the promoter-regulatory sequence and by virus-specific *trans*-acting components. *J. Virol.* **55**:431–441.

55. **Stinski, M. F., D. R. Thomson, R. M. Stenberg, and L. C. Goldstein.** 1983. Organization and expression of the immediate-early genes of human cytomegalovirus. *J. Virol.* **46**:1–14.

56. **Thomson, D. R., R. M. Stenberg, W. F. Goins, and M. F. Stinski.** 1984. Promoter-regulatory region of the major immediate-early gene of human cytomegalovirus. *Proc. Natl. Acad. Sci. USA* **81**:659–663.

57. **Wathen, M. W., and M. F. Stinski.** 1982. Temporal patterns of human cytomegalovirus transcription: mapping the viral RNAs synthesized at immediate-early, early, and late times after infection. *J. Virol.* **41**:462–477.

58. **Weston, J.** 1988. An enhancer element in the short unique region of human cytomegalovirus regulates the production of a group of abundant immediate-early transcripts. *Virology* **162**:406–416.

59. **Wiederrecht, G., D. Seto, and C. S. Parker.** 1988. Isolation of the gene encoding the *S. cerevisiae* shock transcription factor. *Cell* **54**:841–853.

60. **Wiley, C A., R. D. Schrier, F. J. Denaro, J. A. Nelson, P. W. Lampert, and M. B. A. Oldstone.** 1986. Localization of cytomegalovirus proteins and genome during fulminant central nervous system infection in an AIDS patient. *J. Neuropathol. Exp. Neurol.* **45**:127–139.

61. **Williams, T., A. Admon, B. Luscher, and R. Tjian.** 1988. Cloning and expression of AP-2, a cell-type-specific transcription factor that activates inducible enhancer elements. *Genes Dev.* **2**:1557–1569.

62. **Yeager, A. S.** 1974. Transfusion-acquired cytomegalovirus infection in newborn infants. *Am. J. Dis. Child.* **128**:478–483.

INDEX